DISTILLING
THE SOUTH

DISTILLING THE SOUTH

A GUIDE TO SOUTHERN CRAFT LIQUORS

and the PEOPLE WHO MAKE THEM

KATHLEEN PURVIS

THE UNIVERSITY OF NORTH CAROLINA PRESS   *Chapel Hill*

This book was published with the assistance of the Luther H. Hodges Sr.
and Luther H. Hodges Jr. Fund of the University of North Carolina Press.

Designed by Sally Fry Scruggs
Set in Miller and Intelo by Tseng Information Systems, Inc.
Manufactured in the United States of America

The University of North Carolina Press has been a member
of the Green Press Initiative since 2003.

Cover illustrations: corn and other liquor ingredients by Sally Fry Scruggs;
whiskey bottle by Roman Poljak/stock.adobe.com.

All interior photographs by the author.

Library of Congress Cataloging-in-Publication Data
Names: Purvis, Kathleen, author.
Title: Distilling the South : a guide to Southern craft liquors and the
people who make them / Kathleen Purvis.
Description: Chapel Hill : The University of North Carolina Press, [2018] | Includes index.
Identifiers: LCCN 2017044543| ISBN 9781469640617 (cloth : alk. paper) |
ISBN 9781469640624 (ebook)
Subjects: LCSH: Microdistilleries—Southern States. | Distillers—Southern States. |
Distilling industries—Southern States. | Liquors—Southern States.
Classification: LCC HD9390.S672 P87 2018 | DDC 338.4/7663500975—dc23
LC record available at https://lccn.loc.gov/2017044543

CONTENTS

DISTILLING THE SOUTH

Introduction

There I was, standing in a lemon grove, surrounded by trees taller than my head, each one lush with green leaves.

I wasn't in Sicily or Southern California. This lemon grove was in a long greenhouse on a steep hillside in northern West Virginia, in the small arm of the state that curls over the top of Virginia just below Maryland. It's a place that gets an average of twenty-one inches of snow a year, and where the January temperatures usually hover around 20 degrees—not the place you'd expect to find trees with fruit we associate with heat and sunshine.

The lemon grove on the hill: from a window in one end of the greenhouse at Bloomery Plantation, you can look out over the Blue Ridge Mountains.

It seems improbable, and if I hadn't seen it for myself, I wouldn't have believed it. The lemons from those trees were destined to be zested by hand, their skins giving up their flavor to make skinny bottles of Italian-style limoncello right there in West Virginia.

I stood at the window in one end of the greenhouse, looking out at steep hills of the Blue Ridge Mountains stretching off in the distance late on a summer afternoon, and marveled at that view, that greenhouse, and what it represents.

Why would someone go to so much trouble to create an Italian-style lemon cordial in what is surely the most unlikely spot on Earth?

I found the lemon grove on the grounds of a rustic farm called Bloomery Plantation in the countryside outside tiny Charles Town in West Virginia, one of the southern states where changing laws and a new mania for craft food have allowed the flourishing of an entirely new business. Well, a very old business, actually, but one that's coming alive in a whole new way.

Tom Kiefer and Linda Losey, the owners of Bloomery Plantation, have a story that is different and yet familiar all over the craft-spirits business today: on a trip to Italy to attend the canonization of Kiefer's great-great-aunt, they bought a bottle of limoncello and became entranced by the vivid flavor.

Returning to America, they tried to find another bottle, but were disappointed by everything they tried. In America, most commercial limoncello is made with lemons that are zested by machine, including too much bitter pith and altering the flavor.

So they found land in West Virginia on the site of what had once been an old ironworks, and opened a farm-based distillery, where they could grow some of their own lemons and create the kind of limoncello they wanted to drink.

"We love an adventure," Losey says. "That's how we approach life, as an adventure. This hadn't been done, so we thought, 'Why not?'"

Eventually, they went on to add a long row of SweetShine cordials, all based on the moonshine that's been made in the South for generations and flavored with ingredients, some grown on the property, that include pumpkin, raspberries, and Hawaiian ginger. Their black walnut liqueur won best nut cordial in the world at the San Francisco World Spirits Competition in 2015. You can go to the farm and visit their tasting bar in a rustic log cabin dating to 1840.

This is the allure of the craft distilling industry, where passion, heritage, and a desire to express yourself through the creation of spirits are driving a new generation of distillers to make everything from gin and vodka to whiskeys, rums, and brandies, even filling barrels and waiting years to make bourbon, all challenging themselves with a simple aim: if they dream it, they can make it. And you can be a part of it. With this book in hand as your guide, you'll get a unique way to tour the South through the experience of craft liquor.

What drives these new distillers? That's what I set out to discover.

SEE FOR YOURSELF

Visiting distilleries will take you on an unusual journey around the South, getting you off the main roads and back into small towns, farms, and historic areas.

You'll witness a new wave of American entrepreneurialism, small but growing fast, where people are staking their lives, careers, and bank accounts to challenge big-name brands that have become household names, behemoths like Jim Beam and Jack Daniel's.

You'll come away knowing a lot more about what you drink and how it's made. And you'll meet passionate people along the way, from young distillers who are trying to go back to the days when distilleries were a part of every town to people who have been driven by love of a drink to put their own mark on it.

You'll hear about the exploding interest in making liquors that are totally locally sourced as well, boosting state farming systems by using corn and grains grown within a few miles or reaching back to old varieties of rye and corn that haven't been used in more than a generation.

Because they use smaller amounts, many craft distilleries are pinning their claims on "grain to glass" or, in a favorite phrase I heard a lot, "farm to flask." Non-GMO corn strains and organically grown fruits are all finding their way into production, often with help from state universities that are pairing with whiskey programs to train distillers and find better agriculture practices. Almost every distiller I met had also teamed with local farms that use the spent grains to feed local pigs and cattle, so the movement is flask to farm as well.

You'll also meet people like yourself on tours. The people drawn to visit American distilleries come from all over the world, all age groups from millennial to baby boomers, and from all walks of life. Some of the most interesting people I've met on my tours are fellow visitors, people drawn to exploring the world through the lens of food. I've seen families in groups, from young adults to grandparents, and batches of friends out to share an experience. I've seen day-trippers and weekenders, bridal showers and bachelor parties, even people on their first dates. At one stop, I shared the tour with two couples from Belgium pushing a baby in a stroller, out to experience a different view of America. Their questions were delightful and enlightening.

To write this book, I spent more than a year developing six liquor trails crossing the eleven southern states that will give you the most generous number of craft distilling experiences. I drove thousands of miles and visited more than fifty distilleries from Virginia to Louisiana.

I saw more copper stills than an old-time revenue agent, stuck my finger in vats of fermenting corn and streams of clear, freshly distilled alcohol pouring from working stills, sampled an astonishing range of creations and walked through row after row of oak barrels stacked in warehouses, basements, and even portable trailers.

I drove deep into the Florida backwoods and found brandy made from tangerines and absinthe made from tropical ingredients that was colored red from hyacinth. I crushed wormwood between my fingers from bundles hanging to dry in the back of a little shop in Middleburg, Virginia, where the products include a French-style green absinthe right out of the Beaux Arts era. I visited a dying farm town in Georgia that's coming back to life as a food destination thanks to a Dutch couple who took over abandoned buildings to build a rum distillery, creating a new stream of tax revenue and attracting new business.

I found people experimenting with innovative ideas like audio aging, where loud music is piped into barrel warehouses to create tiny waves inside the barrels and hurry contact—theoretically, anyway—with the wood. I met a distiller in Alabama who's putting his barrels of single-malt whiskey into a restaurant-size, walk-in refrigeration unit, raising and lowering the temperature to mimic years of winter and summer in a matter of months.

In Kennesaw, Georgia, I spent an afternoon at a distillery where they're making whiskey in a barn that dates to the 1830s and was one of the only wooden structures in the state that wasn't burned by General Sherman's troops during the Civil War.

DRIVING MYSELF TO DRINK

How did I end up devoting a year to such a crazy endeavor? As a newspaper food editor at the *Charlotte Observer* in North Carolina, I've spent more than twenty-five years covering food in the South, telling the story of my home through the filter of culinary history.

Born in Georgia and raised in North Carolina and Florida, I come from a family that has southern roots that go back generations. If I had my Scots-Irish genes mapped, I don't doubt my DNA would show traces of cornbread, barbecue, and, yes, homemade hooch. One of my grandfathers made bathtub gin to get through Prohibition. When my father was serving with the 1st Marine Division in the South Pacific during World War II, he and his buddies set up small makeshift stills and scrounged supplies to make a rough form of what he fondly called "jungle juice."

The path that led to this project started much earlier, though. About ten years ago, I was approached by Elaine Maisner, executive editor for the University of North Carolina Press, about an idea she had for a unique project:

a series of small cookbooks, each by a different writer and each focused on a single ingredient or occasion in southern cooking.

I ended up writing a book on pecans that was one of the first in the book series called Savor the South. I had so much fun that Elaine asked me to tackle a second book for the series. I talked her into letting me do a cookbook on bourbon.

As a food writer, my favorite stories start with a simple question: Why? The stories that I become the most passionate about usually start with my own curiosity. My husband and I have been classic cocktail fans since long before the cocktail renaissance that took off in the twenty-first century. We're the kind of people who collect books of obscure old recipes and decorate our living room with shelves of antique cocktail shakers that we put to good use.

As a fan of bourbon and bourbon-based drinks, though, I had one simple question: Why is bourbon southern? It's made from corn, after all, and corn is grown all over the country. To answer my own question, I traveled through Kentucky to visit the big distilleries, places like Heaven Hill, Buffalo Trace, Woodford Reserve, and Maker's Mark.

I learned how bourbon is made and, more importantly, why so much of it is made there. While bourbon isn't necessarily southern and there are great bourbons made all over the country, even in New York State, the making of bourbons and whiskeys are intrinsically intertwined with the history of the southern states. I learned about the combination of spectacularly clear water filtered through the region's white marble limestone and the corn that grows so well in the area, and how those conditions led Scots-Irish settlers to discover that the southern mountains were the perfect place to make corn whiskey using the same techniques they had used in their home countries to make barley-based whiskeys, and how that led, eventually, to the development of the style of whiskey we now know as bourbon.

I found myself becoming a little cynical, too. Big whiskey is a big business, and the tours at large distilleries are slick operations that always end in the gift shop, where a couple of samples of their high-proof wares will leave you all too ready to open your wallet for souvenir bottles and T-shirts.

The lectures on those tours are usually colorful, but not all that you hear turns out to be true. The stories and history of American bourbon are as enhanced and mythologized as the tall tales that always seem to bubble up around an open bottle.

Excellent books, such as Robert Moss's *Southern Spirits: Four Hundred*

Years of Drinking in the American South and Fred Minnick's *Bourbon: The Rise, Fall, and Rebirth of an American Whiskey*, have detailed how those colorful stories are sometimes used to paper over unsavory history that involved patent medicines, poor-quality whiskeys dumped on Native Americans, and enslaved workers who did most of the backbreaking work of early distilling.

Still, I loved every minute of my bourbon project, creating dozens of recipes using bourbon in cooking and becoming well versed in the lore and truth of my favorite liquor.

A couple of years later, Elaine approached me again. Always a keen observer of food trends, she had started to notice the sudden growth around North Carolina of small craft distilleries, including one right up the street from her office in Chapel Hill, Top of the Hill Distillery, the maker of TOPO Organic Spirits.

There was something there, she thought. How about tackling a book on that? At first, I wasn't sure. (I'll be honest: my first response was, "It's an awful lot of moonshine, Elaine. And I hate moonshine.") We kept tossing around the idea, though, until we finally realized that the growing distillery industry, even beyond North Carolina, might make a great idea for a travel book. And the idea for *Distilling the South* was born.

Once again, my question was "why"? The growth of the craft-beer industry answered a need: America, for many years, had lousy domestic beer. Craft beer filled a need for something that mostly didn't exist in our food system.

But American liquor has been among the best in the world for generations. So why are people being driven to put everything they have into creating liquor businesses that may never be large enough to compete with large distillers?

The answer to that is the whole motivation behind what we now call craft distilling. That's what this book will help you figure out.

HOW TO USE THIS BOOK: THE SIX LIQUOR TRAILS

Before we hit the road, let me explain how the book is set up: with an estimated 260 craft distilleries in eleven southern states in 2016 and more on the way, it's obviously impossible for you—or me—to visit them all.

For each of the six regional trails I created—and which you can explore for yourself, using this book—I looked for anywhere from six to twelve of the most significant stops. Each was chosen for one of a couple of reasons:

because what they're making or how they're making it made them the most interesting in their region, or because the location made it most likely (or in some cases, less likely) that you'd hear of them while visiting an area.

I also looked for something else: the story of craft distilling isn't just about making alcohol. Like all great tales, it's a story about people. I tried, as often as possible, to find people with intriguing or compelling stories behind what they do and why they do it. I tried to find distilleries that are microcosms of the regions where they operate.

Each trail focuses on two or three states that are grouped together as regions. (Because there's so much there, I broke the Carolinas into two trails for your convenience.) The distilleries I highlighted are arranged in geographic order, usually starting with a logical entry point into the region and then proceeding from south to north and from east to west.

Since this is a travel book that often takes you to small towns or rural areas, I also included occasional traveler's notes on places to stay or interesting places I found to eat or visit nearby.

At the end of each chapter, you'll find a list of every place that I could confirm in that area, with a little information on visiting. That way, if you're in an area and one of the distilleries I've highlighted isn't open or isn't interesting to you, you can find another place to seek out. (One note: state regulations and details on the distilleries are as accurate as I can make them. But rules are changing, and distilleries are opening and closing quickly. Some information may have changed by the time you visit.)

Although I do include notes on what each place makes and what you can taste, I didn't focus on evaluating their alcohol. These are small, startup businesses, after all, and quality in liquor takes time. Besides, in alcohol, quality is in the palate of the taster: you may absolutely love an apple-pie moonshine or a blueberry vodka that might not appeal to me, or I might adore a particular single-malt whiskey or a spicy rye that doesn't taste good to you. Like wine, we all have our own tastes and preferences. If I found something really special, though, I pointed it out so you don't miss it.

In addition, I included a few recipes with each trail. Many small distilleries depend on sales to visitors to supplement what they sell through their distribution to retail stores and bars. You'll inevitably find yourself coming home with an eclectic selection of bottles. That's part of the fun of visiting: small distilleries have more freedom to express their creativity in small-batch creations, and you'll encounter things you won't find in your average package store.

Tips for Touring: Watch Your Step

Like the headache that comes with drinking too much, touring distilleries does come with a few hazards. This is a developing business, after all, so it's not like touring Disney. That's part of its charm: the places you'll visit are operating businesses where you can see what they make up close in a way that means you'll always carry the memories with you.

Before you hit the road, here are a few things to expect:

 Old buildings. You'll find craft distilleries in all kinds of settings, but the most common are in small industrial parks on the outskirts of towns, on farms that might be off the beaten path, and in repurposed stores and factories that were often abandoned in small downtowns. The last category is my favorite: a lot of lovely old brick buildings that once were furniture factories or early department stores are finding new life as distilleries and breweries, bringing new traffic to their towns. Those buildings weren't built for ease, however. While they usually are handicapped accessible, expect steep stairways, drainage grates (skip the heels), and uneven floors. Good footwear is a must, no matter where you go.

 Heat and cold. One reason so much great liquor is made in the South is because the temperate winters and hot, humid summers help alcohol take on color and flavor from wood barrels. Because of that, and because distilleries are basically small factories, most don't have climate control except in the tasting room (and sometimes not even there). Dress accordingly. My collection of folding fans has often made me the envy of tour groups in the middle of summer. I wore out a couple on my journeys.

 On the same note, there are a few areas, particularly in Virginia and West Virginia, where distilleries close or reduce their hours between November and April. Before you plan a stop, check the website or call to confirm they're open.

 Dicey directions. Some distilleries are out in the country or on the back sides of small towns. While Siri is my copilot and my trusty navigation app leads my way, I have found myself in occasional dead zones, or following directions that turned out to be flat wrong. Once, on the edge of the Shenandoah National Forest in Virginia, all cell-phone coverage disappeared completely. I ended up in a small coffee shop owned by several women who quickly whipped out paper atlases, put their heads together to decide on my best route, and sold me the best slice of quiche I've ever eaten. Serendipity, as always, is the best part of travel. Still, a "belt-and-suspenders" approach is wise. Keep an old-technology paper map handy as a backup.

Build in travel time. When you're trying to visit more than one distillery in an area, you may not always find them open on days or at times that are the most convenient for your schedule. If you really are interested in seeing a particular place, it's worth checking, though. Many distilleries will cooperate if you give them a little notice.

 If you're planning a tour with a large group, make sure the place can accommodate you. In urban areas, ask about parking, particularly for tour buses.

 Mansplaining. How can I put this delicately? There are certainly many women, like myself, who appreciate well-made spirits, and there are many women working in the industry as well. But the distillery world, both in those making liquor and those visiting to see it, is still very male. I have encountered occasional distillers who talked down to me—or answered all my questions by turning to my husband—and male visitors who felt the need to "translate" for me. (These were the exception, rather than the rule, I'm happy to report.) If you're a woman making visits, please don't be intimidated or let that stop you from visiting and asking lots of questions. No one starts out knowing everything about how alcohol is made. That's why there are tours, so we can all learn.

 Budget accordingly. The rules that cover what you can do, how much you can buy, and how much you can taste vary from state to state. Prices for tours range from free, including a tasting, to Georgia, where the only way to buy a bottle on-site is to pay for it as a "souvenir," making your tour $30 or $40. On average, most tours are $5 to $10, which usually includes the tasting.

 Buy when you can. While larger distilleries may have established distribution far beyond their state, a lot of small places are offering tours to develop potential customers. You're not required to buy anything, but it's a good way to support a business when you find something you really like. One warning, though: because these are small businesses, they don't have the economy of scale. Bottles at most distilleries usually start at $30 and go up, sometimes way up, to $50 or $60 for aged spirits. Just because you're at the source doesn't mean you'll get a bargain.

However, that means you can find yourself wondering what you're going to do with the products you bring home. So I reached out to a few cocktail experts and to the distillers themselves for ideas on how to use some of your finds.

Now, it's true that some young distilleries are pouring samples that aren't necessarily ready for prime time. Give them time: the quality, quantity, and character of what they're making may be very different in just a few years. That's part of the fun of getting to know them now, as a new industry begins to grow.

Getting Crafty

The Basics of the Distilling World

Breweries make beer, distilleries make spirits. At Asheville Distilling, you can tour Troy & Sons whiskey and Highland Brewing right next door.

The first thing you have to tackle in creating liquor trails to craft distilleries is answering one big question: what's a craft distillery?

In deciding which distilleries to include in this book, I defined them this way: I was looking for distilleries that have small outputs and hands-on operations, and that are the creative expression of a small group of owners without corporate ownership. While the number of years in operation isn't a reliable criteria for addressing craft, many of the distilleries that fit in the category have been in business for less than fifteen years, mostly because the changes in regulations that allow their existence are relatively recent in many states.

There's plenty of room for argument, though. In the food world for the last few years, the words "craft" and its close cousin, "artisan," have become code words, terms that pack in a lot of vague ideas with no clear definitions. What do we mean when we say something is "craft"?

In Charleston in 2016, I attended the first meeting of BevCon, a conference for the craft beverage industry that brought together hundreds of makers of liquors, wines, beers, and sodas, along with bartenders, hobby mixologists, and beverage writers.

The opening session, led by *Tasting Whiskey* author Lew Bryson, tackled the issue head on: "What Is Craft?"

Bryson and his panel, including Lance Winters of St. George Spirits in

California, Paul Hletko of FEW Spirits in Illinois, and Scott Blackwell of High Wire Distilling Co. in South Carolina, debated all the ways we define craft businesses: Is it based on size, or ingredients, or age? Is it based on sustainability or localness? Is it based on perceptions of quality or honesty? Is it handmade or does it imply a minimum of automation?

Hletko, chair of the American Craft Spirits Association (ACSA), admitted that even using the word in their name can be a problem because the term is so emotional and impossible to define.

"It's something that stirs the creator," said Blackwell. "It's creative, it's his artistic expression. Craft is about the action, not the words."

That all may be true, but any master distiller, whether working for a tiny, 500-barrel distillery or a giant like Jim Beam, would certainly claim to be stirred by artistic expression.

Even inside the industry, the definitions can be vague. The American Craft Spirits Association says "craft spirits" are distilled spirits made by licensed producers that have no more than 750,000 proof gallons, or 394,317 nine-liter cases, removed from bond (or released on the market), that market themselves as craft, and that aren't openly controlled by a large supplier.

I can think of more than a few questions about that definition, including how simply marketing yourself as craft makes you craft, and what "openly controlled" means.

OK, so what about the American Distilling Institute? They define craft as products of an independently owned distillery with maximum annual sales of 52,000 cases, where the product is physically distilled and bottled on site.

Sounds a bit clearer, but in my travels, I found plenty of distilleries that certainly are craft but were having at least some of their products made elsewhere while they got up to speed. Does that mean they don't count? And where does that leave places like Bloomery Plantation, which doesn't make its own moonshine but uses it as the base for things they create and flavor with at least some ingredients they grow themselves? Purchasing whiskey that was made elsewhere to create your own blend isn't new, after all. It's a practice that dates to the nineteenth century.

Hletko says his group tries to be as inclusive as it can, based on an ethics policy that governs its members.

"We want everybody," he says. "We want people who are growing their own grain and distilling it from scratch, but we want people who buy grain and distill, and people who buy moonshine and turn it into cordials, and somebody who goes out and carefully sources barrels [of products].

"The craft in all these things is admirable, and we at ACSA want to be a part of all that."

In some places, such as North and South Carolina, pretty much any legally operated distillery qualifies as a craft distillery because the business is so new. In states that have long histories of illegal distilling, legal distilling is a very young animal. All over the mountain South, which has a long tradition of moonshine (and yes, I did eventually find a few moonshines that even I liked), I met plenty of people who would answer my question about how long they had been distilling by saying, "Well, *legally*, for about ten years." The implication was clear.

In places like Tennessee and Kentucky, where distilling has been going on legally since the late 1700s—with the exception of that little hiccup known as Prohibition—craft distillers stand out because they're upstarts, far smaller and more able to experiment than the big players around them that produce millions of barrels a year.

HEADING SOUTH FOR FREEDOM

While eleven southern states accounted for only about 28 percent of the 1,300 or so craft distilleries operating nationwide in 2016, according to a study commissioned by the ACSA, the South itself plays a much bigger role in the history of American distilling. Knowing a little bit of that history before you start touring will help put it in perspective.

Brace yourself, because this is history in very broad strokes: entire books can (and have) been written about liquor in the South. So let's just draw a sketch.

In the beginning, and by that I mean the colonization of America—no disrespect, but at least along the East Coast, distilling was one thing native peoples hadn't already figured out long before the Europeans arrived— the eastern part of what became the United States was a place full of farms settled by people who knew how to make a living off what they could grow. If you're going to feed your family through the winter, you'd better make the most of what you can grow in the summer, especially in years when you have a banner harvest.

If you grow fruit trees, you can eat the fruit, dry the fruit to cook later, and turn some of that fruit, conveniently loaded with its own sugar, into brandy. If you grow grains, you can eat the grain, grind it into flours and meals to make other things, use the grain to feed your livestock—and, with the help of a pot still, turn it into whiskey. Handy system, right?

Many of the people who came from Europe as settlers were from the British Isles, particularly Ireland and Scotland—both places where the practice of distilling grains into spirits was already well known. They generally worked with barley, but they knew how to make a grain mash, use yeast to convert its starches into sugar, ferment the sugar to produce alcohol, and then heat the liquid from the mash to separate the alcohol from the water. (It's a little more complicated than that, but not much more.)

Here on a new continent, there was plenty of land to create farms. Leaving out the imports, such as rums made on sugar plantations in the Caribbean and specialized wines like madeira, it didn't take long for farmers to figure out that maize, that lovely corn that grows so well in all kinds of soils, could be treated much the same way as cereal grains, giving the new Americans a handy source for spirits. And from Pennsylvania into the South, rye grew well and was better suited to the climate than barley. Virginia, particularly, is still known for rye-based whiskey.

So those two things, brandy and whiskey, became lucrative ways for farms to make something they could store for their own use or trade and sell.

Now, jump forward a little, past the American Revolution and the forming of a new nation. By 1791, our first president, George Washington, had a problem: the new government was saddled with a lot of debt from fighting that war. So Washington and his treasury secretary, Alexander Hamilton (yes, him—star of stage and Broadway musical), decided to levy an excise tax on distilled spirits, particularly corn and rye whiskeys.

That didn't go over very well with farmers along the outer edges of the new republic, from Maryland to South Carolina and especially in Pennsylvania. Many members of the new nation were still prickly about that whole notion of taxation and representation.

Violence broke out in Pennsylvania between tax collectors and farmers who refused to pay the tax. In what has become known as the Whiskey Rebellion, Washington was able to use his military to stop it, proving that the new government had the strength to enforce its own laws.

But the new tax was very unpopular. When Thomas Jefferson was elected president in 1801, the whiskey tax was repealed.

It was hard to collect, anyway. The expansion of settlers into the area that eventually became Kentucky had started in the 1760s with Daniel Boone. Both Scots-Irish and German pioneers brought distilling along with them into an area that was isolated from the more settled coastal areas by the rugged terrain of the Appalachian Mountains. In one of history's happy accidents, they also found a place with the perfect conditions to make good corn whiskey, including pure, iron-free water. The location, along the Ohio River, meant they could load barrels of whiskey onto flatboats and ship it by river, starting at the Ohio and eventually to the Mississippi, finally ending up in the lucrative market of New Orleans.

Today, long-overdue credit is being given to enslaved distillers in the region, who were often the people who did the hot, hard work of running stills. A slave who had the knowledge of distilling was a valuable person, and some historians believe that some may have brought the knowledge of making fruit-based spirits from West Africa. Some distilleries, such as Jack Daniel's and George Washington's Distillery at Mount Vernon, are starting to include the role of black whiskey makers in their histories.

By the late 1700s, distillers in Bourbon County, Kentucky, once part of a vast territory that stretched back to Virginia, discovered that charring, or burning, the insides of the barrels before they filled them with corn whiskey gave their clear, fiery whiskey a beautiful reddish color and a smoother, more mellow flavor during the months it took to fill a flatboat and navigate the long trip by river. Pretty soon, the barrels stamped with their county of origin, Bourbon, developed a reputation as something special.

Meanwhile, in the rest of the South, distilling was still going on as well, often in remote areas where people were more likely to keep their liquors local, either for personal use or for sale. Take a look at the map of the South and the ACSA's numbers for craft distilleries in each state as of 2016 and you can see the legacy of that history: if you start in the upper northeast corner with Virginia, there were thirty-five distilleries listed. That keeps on in the nearby states: forty-two in North Carolina, twenty-nine in South Carolina, thirty-two in Tennessee, and thirty-five in Kentucky. (Those numbers don't include the large, well-established distilleries in Kentucky and Tennessee.)

Once you get away from the mountains and move deeper into the Bible Belt, where regulations and social norms were more restrictive, the numbers drop off quickly: seventeen distilleries in Georgia, eight in Alabama, three in Mississippi, and eleven in Louisiana.

Still, in the South, making alcohol was a way of life, and in many places, it stayed that way for more than one hundred years—until Prohibition upset the whole apple cart, brandy and all, in the early years of the twentieth century, almost destroying the American distilling industry.

Stills and the Basics of Distilling

Distilleries are definitely not places where you can say, "Seen one, seen them all." Yes, the basics are the same, but every distillery is set up a little differently. Touring more than one is a good way to learn about all the ways you can tackle the job of turning raw products into hooch.

You can also learn a lot by comparing the explanations of tour guides and distillers. Each one will tell a slightly different story or highlight a different part of the process. So what do they all have in common?

One thing every distillery has to have is a still. The still is the heart of the operation and one of the biggest investments a prospective distiller has to make.

Makers love their stills so much, many even name them. You'll see everything from a squatty boiler nicknamed "Cartman" from *South Park* (because "it's short, fat, and loud") at Corsair Distillery in Bowling Green, Kentucky, to stills named for the owner's politics ("Liberty" and "Democracy" at Muddy River Distillery in Belmont, North Carolina). At Copper & Kings, the brandy maker in Louisville, the three stills are named for women in the lyrics of a particular Bob Dylan album, in keeping with the brand's music theme.

You'll read more about them in my tour descriptions, but it helps first to get a little idea of what stills are and how they work.

Stills don't have to be made from copper.

You'll see some that are stainless steel or a combination of copper and steel. But the majority are copper, and they are beautiful: lustrous, glowing, with tops in a variety of shapes, from an elongated cap sometimes called a horsehead to swirls that look a little like onion domes.

Just as finding good water sources was critical to the story of southern alcohol, using copper is another happy accident of history. For hundreds of years, stills were made from copper for two reasons: it's a soft metal, so it's easy to shape, and it conducts heat beautifully, able to get hot and cool off many times before it cracks.

Thanks to molecular science, though, we now know that copper turns out to have another magical property: certain things, particularly sulphur, yeast, and bacteria, cling to it. When you run the liquid from a grain mash through it, it removes things that could give your liquor off-flavors, and it helps to create alcohol with a smoother taste.

All stills work on a simple principle: water boils at 212 degrees (at sea level, anyway). But alcohol boils at a lower temperature. So when you heat a mix of alcohol and water, the alcohol gets hot first, rising as steam before the water does. There are actually numerous compounds in it, so as the heat rises, you get three types of alcohol.

Heads (including foreshots). At 133 degrees, the first alcohol molecules that rise include acetone, methanol and "hydes," such as formaldehyde. Most women will recognize the smell of acetone: nail polish remover. All of these are poisonous in various degrees, and gave illegal moonshine the reputation of "drink that and you'll go blind." When you're distilling, you have to remove that from the alcohol you want to drink. (Most small distillers, trying to make every dollar count, save the heads to use as cleaners. Even distilleries in barns and old buildings are usually squeaky clean.) While there is equipment for taking readings of the alcohol content, many distillers simply watch the temperature or use their nose and taste buds to decide when the run moves from the heads to the next stage.

Hearts. At 173 degrees, the alcohol molecules that rise are ethanol, also known as neutral grain spirits, the good, clean alcohol that is the base for every liquor. Depending on the size of the still, this may run from the still for a couple of hours.

Tails. At 206 degrees, the last of the run starts to change in flavor. Fusel oils, or flavor compounds, in the last bit of alcohol are usable, but they also can be harsh or unpleasant, usually described as having a "stinky feet" smell. Some distillers toss them out and will brag about doing so. Others save them to add to the next run, sort of like a sourdough starter, and they'll brag that they're adding something that ends up being double- or triple-distilled. There's a case to be made for either approach.

Simply boiling the mash and catching the molecules of alcohol isn't all there is to it, though. You've created alcohol steam, and now you have to turn that steam back into liquid. The basic way to do it is the worm: a copper coil surrounded by something cold, from well water to mechanically chilled water. The hot alcohol steam rises up from the still, then into the copper coil, where the cold around it turns it from steam to liquid, kind of the way evaporating water rises from an ocean to form a cloud and then falls again as rain.

What makes all this possible is the still. One of the things that makes touring so much fun

You'll see pot stills, column stills, and hybrids of the two, which allow small distilleries to maximize which styles of liquor they can make.

is the variety of stills. There are capsule styles and upright styles, and even stills with glass ports etched with catchy names or the faces of ancestors who were moonshiners. That customization is part of what makes craft distilleries different from mainstream ones.

In general, though, stills fall into three types.

Pot stills. Squatty, round, and sometimes shaped like an acorn, the pot still is the oldest style. When you put the mash of grain and liquid into it and heat it up—originally with a log fire underneath, although most places now use much safer steam jackets—the steam rises to the top and into a pipe that leads to the worm, leaving the water and spent grain behind.

Column stills. Tall and rising as high as you have room, a column still looks like a cross between a big flute and the Beatles' yellow submarine, with port windows set at intervals all the way up. Behind each window is a set of plates, usually flat with holes, although some are shaped like bells. The steam of water and alcohol rises up, catching on the holes in the plates, raining back down and rising again. By the time the molecules get to the top and cross over to the next stage, you only have alcohol, with the water left behind.

Hybrid stills. A third category combines a pot still and a column still, either side by side or in the same construction. A custom-built hybrid can be very expensive—you don't pick them up on the used-still lot. But it can have advantages for a young distillery. It combines the best of both forms, and it allows for more customization in what you create, increasing your options as you grow and branch into more liquors.

One thing slowing the growth of distilleries is the difficulty in getting a still built. It can take years (and thousands of dollars) on the waiting list now to get a still from popular makers such as Vendome in Kentucky.

UNCORKING CRAFT DISTILLING

The growth of the modern craft spirits industry is nothing short of astonishing. According to "The Craft Spirits Data Project," a study commissioned by the ACSA in 2016, there were 204 craft, or small, distilleries nationwide in 2010. By 2015, that had jumped to 1,163, and to 1,315 a year later. If that growth rate continues, the number will more than double, to 2,800, by 2020.

While market share for the sales of craft spirits accounts for only 2.2 percent of the volume and 3 percent of the value of American liquor sales, the growth rate is already impressive: that 3 percent in value has grown from only 1.1 percent in 2010, and the growth rate climbed 27.9 percent in value in those five years.

In a way, that mirrors another American phenomenon of the last decade, the craft beer industry. We've become a nation of people who may buy fewer alcoholic beverages than we used to, but we're buying better, spending more money on fewer things that are better made.

While the equipment is similar, though, a brewery is a simpler thing to build than a distillery. And with a brewery, you have a product you can sell very quickly, in a matter of months.

Liquor is different. The beginning of the distilling process, if you're making whiskey, anyway, involves turning grains, especially corn, into a mash and letting it ferment. The sour-tasting stuff you get is called distiller's beer or low wines.

Next, you have to distill it, or separate the alcohol from the water, using some form of a pot or column still. Then you have more work ahead of you. Gin, vodka, and corn whiskey (by whatever name you want to call it, including white dog, white lightning, or moonshine) can go straight into a bottle and be sold right away. That's why so many distilleries start out with clear-liquor products. But if you want to make something that you age, such as single-malt whiskeys, bourbons, and some rums, you've got to get your hands on barrels, which have become very expensive; you've got to find a place to store them while they age; and you have to wait, sometimes for years, while what you put in the barrel becomes something you can sell for the highest price possible.

The investment is spectacular. At Call Family Distillers in Wilkesboro, North Carolina, where the tradition of making moonshine goes back seven generations—illegally, of course—owner Brian Call told me about creating a legal distillery as a family business.

"I thought it would cost me $30,000 or $40,000," he said. "And then I saw a million dollars in my rearview mirror and I was still headed north."

Obviously, though, one reason this business is growing so fast is because there is money to be made at it. In 2015, according to "The Craft Spirits Data Project," the entire U.S. craft market produced 4.9 million cases and $2.4 billion in retail sales. Of course, a split is already forming, with the largest craft distilleries, the 2 percent that have enough output and sales staff to reach into out-of-state sales, accounting for 60 percent of all cases sold. The smaller places, about 91.7 percent of the craft distillers, account for 12.7 percent of the sales. For them, direct sales—those of us who visit and buy a bottle—are 25 percent of their business.

Part of what's driving so many people to stake their financial futures on distilling is the sheer romance of it, the desire to make something that you can truly call your own.

Paul Hletko of ACSA is a maker of spirits himself, the owner of FEW Spirits in Illinois.

"I think people get really motivated by what they consume and what they share with friends." He uses the example of a friend who makes vacuum cleaners: "No one cares. I would wager that the last time you talked about the new vacuum cleaner you bought was exactly never. That conversation doesn't ever happen.

"But at least once a week, I'll sit around with friends and we'll talk about the spirits we're drinking or the beer we're having. The relationship people have with their food and beverages is very different, and that passion carries through.

"We could make a lot more money selling widgets than booze, but it's not nearly as much fun."

Over and over at distilleries, when you ask people about how they got into the business, you'll hear variations of similar stories: I had made my living in medical technology, or selling insurance, or working for a big corporation. And rather than retire, I decided to spend the rest of my working life doing something more fun.

Or: I found out that my great-great-grandfather was a distiller. So I wanted to bring the family business back to life.

Or: It was 2008, I had the chance to take a buyout, so I decided I'd try my hand at making gin, making rum, making brandy.

And then there's this one that several distillers told me: I was sitting around playing poker with friends and drinking a little whiskey, and I thought, I wonder if I could make this?

In the same way that the economic downturn before 2010 led so many young people to create food businesses making charcuterie or cheese or to open whole-grain bakeries, that financial bump in the road steered a lot of people toward the idea of creating spirits they could call their own.

When you visit these small distilleries, you're seeing a new form of the American dream: the dream of small-scale manufacturing, of making something with more meaning.

Know Your Alcohols

Because of the well-organized bourbon trails in Kentucky, it's easy to get the idea that all craft distilleries in the South are making whiskey. Actually, there's a long list of liquors you'll find out there, including rare and even vanished styles that are being rediscovered. Here are the main kinds and their definitions.

ABSINTHE

A high-alcohol distilled spirit flavored with herbs and dried flowers, particularly worm-wood, fennel, and anise. (The name comes from absinthium, the Latin word for "worm-wood.") While green is traditional, it can be clear or red, too. It's often called a liqueur, but it usually doesn't have added sugar. Its reputation as a hallucinogen led it to be banned in the United States and most areas of Europe through most of the twentieth century. That has now been disproven, and absinthe is enjoying a renaissance, particularly in the craft cocktail world.

AMARO

An Italian-style liqueur that's usually both sweet and bitter (amaro is Italian for "bitter"). Because there's no definition for what it has to contain, many recipes date to ancient monasteries and there is a wide range of flavors. In Italy, it's served as a digestive aid after a meal. Amari used to be rare in America, but they've been discovered by cocktail fans and craft bartenders, and there are versions being made by a few distilleries, including High Wire in Charleston.

BRANDY

All brandies are based on fruit. Grapes, apples, and peaches are the most common, although pears and even persimmons can be used to make brandy. The fruit is crushed or turned into juice before being fermented and distilled.

CORDIALS AND LIQUEURS

Both are usually sweetened distilled liquors, usually at least 2.5 percent sugar. Liqueurs are often very high in alcohol and are sipped neat in small quantities or used with other liquors to make cocktails. The term "cordial" can get confusing: in Britain, cordials are sweet, nonalcoholic syrups, but in America, a commercially made cordial almost always has alcohol. Cordials can be flavored with anything from fruit, herbs, or spices to cream, chocolate, coffee, or nuts.

GINS

Based on a neutral grain spirit, gin is then returned to a still and infused with aromatics, usually through steam or vapor. By definition, the predominate flavor has to be juniper, but the amount can vary, with some gins being "juniper forward" while others are specifically made to tone down the juniper. The range of other aromatic ingredients can cover a long list, from citrus peels to lavender, or even Szechuan peppercorns. Many craft distilleries are starting to age or "barrel rest" gins, usually for less than a year, to add color and flavor variations, ending up with something that fits better into whiskey cocktails.

American craft gins fall into a number of styles, and you'll hear all kinds of phrases to describe them.

American-style. Sometimes called New West or "botanical style," it can be "anything goes," based on the flavor profile the maker wants to reach. It can have as many as twenty added flavoring agents.

London dry. A clean, clear version where no single flavor dominates (except the juniper—it's often very piney). At some tasting rooms, you'll hear it described as "Beefeater-style" because of the familiar brand your parents used to make martinis.

Navy-strength. It's 57 percent alcohol by volume compared to 40 to 45 percent for regular dry gins. There are many stories about the name. Some say that when the British navy carried gin on ships and each sailor got a daily allotment, a stronger gin would still allow gunpowder to ignite if the gunpowder got splashed with gin in the hold. Others say it was because the ships' purveyors would make sure the gin they were buying wasn't watered down by mixing it with a few grains of gunpowder and exposing it to sunlight with a magnifying glass. If the gunpowder didn't light, the gin was too high in water.

Old Tom. An old style that is finding a revival in the craft-cocktail world, it is a slightly sweetened gin. However, there is no standard definition, so Old Toms can range from clear and sugar-sweetened to barely sweetened and reddish due to aging in wine barrels.

RUMS

A spirit distilled from sugarcane or from something made from sugarcane, usually cane syrup or molasses. After you boil sugarcane juice until it forms crystals of sugar, the dark, thick liquid left behind is molasses. Molasses or syrup are mixed with water and yeast and allowed to ferment, making a wine that's distilled to separate the alcohol. It's made in much the same way that whiskey is made, distilled in a pot still or column still, although there are some differences in the process. Rums can be aged in charred oak barrels to add color and flavor. Beyond that, there are several popular styles:

White rum. Clear rum that is usually unaged, although some white rums are barreled and filtered.

Golden rum. It can be barrel-rested for a short time, adding a little color, although some versions get color from the addition of caramel.

Spiced rum. There's no rule on how these can be flavored, but the mix usually includes cinnamon, nutmeg, and vanilla. Caramel can be added for color and flavor.

Rhum agricole (say it *rum ah-gree-COAL*). It's made from fresh sugarcane juice before it's boiled to make cane syrup or molasses. It's popular with craft distillers who either grow their own sugarcane or have it grown locally. Because cane juice oxidizes quickly, it can only be made during the harvest season. It's prized for having more flavor notes and more vegetal characteristics.

VODKA

While they wait for more complex products like whiskeys to mature in barrels, many distilleries start out with vodka. By definition, it's a neutral spirit, and it can be made from anything that will ferment, from grains (rye or wheat) to root vegetables to fruit or cane syrup, even muscadine grapes. There are corn-based and rice-based vodkas, too. Because the flavor is so neutral, you'll see a lot of flavored versions as well.

If it's distilled from grains or corn, it's whiskey. But within that definition, there are a lot of types.

Bourbon. The simplest definition is that it is a whiskey distilled from a mash that is at least 51 percent corn (and up to 100 percent) and can include malt and wheat or rye. The full definition was set by Congress in 1964 (with a lot of influence from the bourbon and barreling industries). Under the Federal Standards of Identity, it also must be made in America (not just Kentucky), distilled at 160 proof or less, put into a new charred oak barrel at 125 proof or less, and bottled at 80 proof or higher. It has to be aged at least two years (for straight bourbon) or four years (if it's less than four, it has to have the age somewhere on the label; the age statement must be the youngest bourbon in the bottle, so it can contain whiskey that is older). It can only be cut with water or other bourbon and can't contain any colors or flavors.

Corn whiskey. Distilled from a mash of at least 80 percent corn (and sometimes 100 percent). It usually is clear and unaged, although some distillers are starting to barrel-rest or barrel-age to add a little color and flavor. It's worth noting here that the term "moonshine," often applied to corn whiskey, can be controversial. "Moonshine" originally was a slang term for any illegal alcohol. But there is no legal definition, and with the rise of legal craft distilling, many unaged corn whiskeys are being called moonshine. At some distilleries, what's sold as moonshine may be made according to handed-down methods and recipes, or it may be neutral grain spirit that's bought and repackaged. It may be flavored and colored with artificial ingredients or with natural ingredients, including spices and fruit.

Irish whiskey. It has to be made in Ireland from a mash of cereal grains (usually barley) with a malt component, either double- or triple-distilled in a pot still to less than 94.8 percent alcohol by volume, and aged at least three years in wooden casks no larger than 185 gallons.

Poitín (also spelled poteen). Pronounced *pah-CHEEN*, it was the Irish version of illegal moonshine. Although it's usually made from malted barley, there's no legal definition of the mash, so it can be made from other grains, potatoes, or even treacle. Since Irish whiskey has to be made in Ireland, some American craft distillers make Irish-style whiskey and call it poteen or the Irish spelling, poitín.

Rye. American rye whiskey follows the same standards of bourbon, except that the majority of the mash must be rye instead of corn. (Canadian rye is a style of whiskey and isn't required to contain rye at all.)

Single malt. This term is getting very confused. Technically, it means a malted-grain whiskey from a single distillery. However, it's starting to be used to refer to whiskey made from only malted barley, or in some cases, malted rye. Some people confuse the term with Scotch, because so many Scotch whiskeys are single malts. But American single malts usually don't have the distinctive earthy flavor associated with smoking barley over peat.

Tennessee whiskey. It must be made in Tennessee and filtered through charcoal made from sugar maple wood. Otherwise, the process, including barrel-aging, is the same as bourbon.

Virginia and West Virginia

LIQUOR TRAIL

1

Virginia and West Virginia

1 George Washington's Distillery, Alexandria, Va.

2 James River Distillery, Richmond, Va.

3 Reservoir Distillery, Richmond, Va.

4 Catoctin Creek, Purcellville, Va.

5 Mt. Defiance Cidery & Distillery, Middleburg, Va.

6 Bloomery Plantation, Charles Town, W.Va.

7 Copper Fox Distillery, Sperryville, Va.

8 Smooth Ambler Spirits, Maxwelton, W.Va.

WEST VIRGINIA

64

Lewisburg

81

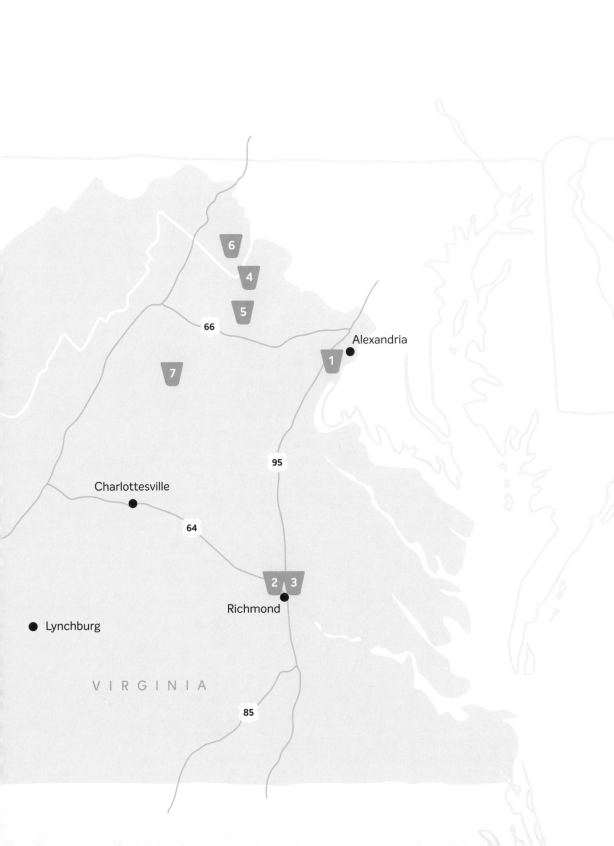

Alexandria

66

95

Charlottesville

64

Lynchburg

2 3

Richmond

VIRGINIA

85

6

4

5

7

1

WITH SO MANY AREAS easy to reach from Washington, D.C., and a breathtaking countryside of hills, mountains, valleys, and forests, touring the distilleries of Virginia and West Virginia is a traveler's dream.

Many times while exploring this region, I wished I was traveling by motorcycle or that I had a GoPro strapped to the roof of my car to capture the scenery: along the highways that parallel the scenic Skyline Drive, steep, green valleys drop away so dramatically that it feels like you're driving along the rim of a bowl made of the Earth itself.

Distilleries in the eastern part of the region are close to Virginia's burgeoning wine industry, making it ideal for day-trippers from D.C. who can head out for a day in the vineyards and stumble on distillery tasting rooms along the way. (Be sure to check the calendar before you head out. Several distilleries in this area are only open from April through October.)

While Virginia has a state-run distribution and retail system, it also allows unlimited sales at distilleries. At Mt. Defiance, a tiny distillery in Middleburg, owner Marc Chretien likes to call his shelf of whiskeys, absinthes, rye, and rum "the smallest ABC store in the state."

In Virginia's ABC system, there are two levels of retail: special order and on-shelf. A ten-person board approves every spirit that's stocked at the 350 state-run stores, no matter where it's made. But while shelf space tends to be protected for the large brands, distilleries also can qualify for special-order sales, meaning that it has to be in the state's system and can be ordered.

That and on-site sales at distilleries gives small makers a chance to find a market, and has helped Virginia quickly become one of the most exciting states in the nation for the development of craft spirits. A couple, including Copper Fox and Catoctin Creek, have developed distribution far beyond the state. You'll spot their labels behind the bars at some of New York's hippest whiskey bars.

Samples served in tasting rooms are limited to three ounces, total, per person. At most distilleries, that means you may have to choose three to five things to sample out of a lineup that might include eight or ten varieties. With many distilleries experimenting with a wide range of products and "one-offs" or single experiments, that can lead to hard choices unless you team up to share sips with a partner. Virginia distilleries also are allowed to serve cocktails, although not all do.

All over Virginia, you'll hear references to "the Virginia style," which means 100 percent rye whiskey. Rye is native to Virginia, although it's hard to grow there now because of the warmer winters attributed to climate change.

Meanwhile, across a border so winding it's hard to tell when you've left Virginia for West Virginia, the growth of distilling has been a windfall in a state that struggles with the economic hardship of coal country. Particularly in the western part of the state, you'll find distilleries, like Smooth Ambler outside Lewisburg, in rugged country that's ideal to combine with outdoor adventures like camping, hiking, and rafting.

Craft distilleries in West Virginia struggled just a few years ago with tax rules that penalized the craft distillers by pitting tasting-room sales against liquor stores. Distilleries used to have to pay 10 percent of their sales to the liquor stores. That's been reduced to 2 percent, easing the burden and helping the distilleries to begin to thrive.

My other trails are arranged in a geographic circle. But in the Virginias, I'm going to take one out of order, so we can start with the early days of distilling in America.

1 GEORGE WASHINGTON'S DISTILLERY

George Washington's Mount Vernon, 5514 Mount Vernon Memorial Highway, Alexandria, Va.; www.mountvernon.org. Closed November through March.

When we last joined George Washington, back in the history section, he was president of a young nation desperate for tax revenue.

Now, skip forward to 1797. Washington retired from the presidency and returned home to his 8,000-acre estate, Mount Vernon, originally built by his father, Augustine, in 1735. The land along the Potomac River included a stone gristmill on Dogue Run, a small creek about two miles from the great house. Along with tobacco, the farm produced wheat, corn, oats, and barley, so the three-story mill was kept busy making flour and meal.

When Washington returned from running the country, he was an older man who was interested in making his farm run as profitably and efficiently

At the Mount Vernon historic site, George Washington's distillery has been re-created on the same spot as the original, near the gristmill fed by Dogue Run.

as possible. He hired a Scotsman, James Anderson, as his farm manager. Anderson knew whiskey-making from his native country, and he noticed that Mount Vernon had a gristmill and plenty of corn. He suggested that Washington add a distillery. At first, Washington resisted—he didn't know anything about the business of whiskey, and he may have been reluctant about the rowdy trade it attracted. While he was known to be a fan of rum and Madeira, he also was a proponent of moderation.

Anderson persisted, though. John Fitzgerald, Washington's friend and former aide de camp, owned a successful rum distillery in Alexandria. He helped Anderson persuade Washington, who finally agreed to let Anderson buy two pot stills.

In the first year, 1797, they produced 660 gallons of corn whiskey and sold it all. They couldn't keep up with the demand, so Washington allowed Anderson to build a full distillery with five pot stills, located just over a rise from the gristmill.

By March 1798, they were making 7,500 gallons of twice-distilled whiskey and selling it for 60 cents a gallon, or $1 a gallon if it was distilled four times.

Aging whiskey in charred oak barrels was just starting to show up in Kentucky; in Virginia, whiskeys might be stored in barrels for a short time, but aging hadn't become a regular practice.

In 1799, the year Washington died, Mount Vernon produced 11,000 gallons, making it the largest distillery in the country.

Sadly, after Washington's death, the distillery didn't last long. His nephew inherited the estate and had a fight with Anderson, who left. The distillery was shut down in 1808, and the building burned to the ground in 1814.

For a long time, Washington's activities as a whiskey maker were almost forgotten. There were plenty of records that showed the distillery existed, but when Mount Vernon was restored as a historic site in the 1920s, Prohibition was still in force. No one wanted to suggest that the father of our country did something as unsavory as make whiskey.

That turned out to be a good thing. Since no attempts were made to reclaim the site, it remained untouched. In 1995, the site was excavated in a massive archeological project, part of a restoration of the Mount Vernon grounds. It was so well-preserved, some of the original brick floor was still there, and the fires under the stills had left enough traces to show exactly where they were.

By 2009, the distillery had been rebuilt and the estate began making whiskey there again, using costumed interpreters to run it.

You don't have to buy a full ticket to Mount Vernon to tour it. Just pull into the Mount Vernon Distillery parking lot two miles before you reach the main estate entrance and buy a $5 ticket in the gift shop. That gets you a tour of both the operating gristmill and the distillery itself. (You can also tour the gristmill and distillery with a full pass to the estate.)

In the gristmill, you'll get an excellent look at stone milling, using water sloughs from pretty little Dogue Run, right outside the building. They still grow corn and tobacco on the estate, although when I was there, deer had eaten all the wheat, so it came from elsewhere. If you have any kids in your group, it's exciting to watch. When the sluice gates are opened and the big stone wheels and wooden gears start to turn, it's like being inside a three-story wooden clock.

The long, stone distillery building is just over a small rise, with an easy path leading to it. They make unaged rye whiskey, true to the style of Washington's day. It's believed to be the only legal distillery in the country still heating pot stills over wood fires.

The distillery fills one room on the bottom floor, making it easy to follow the steps they would have taken to make whiskey by hand: working with a long-handled scoop fashioned from a wooden pole with a small wooden bucket, like a small barrel, on one end, they would have poured 30 gallons of boiling water over the ground corn, then stirred it while it cooked to get the starch out. More boiling water would be added, followed by the rye. Letting that stand overnight, they'd then add malted barley to start breaking down the starch into sugar. After adding yeast, it would stand for three to five days, while the yeast ate the sugar and turned it into alcohol.

All that would be moved to a pot still for a stripping run to produce 40-proof alcohol. After cleaning out the spent mash (in Washington's day, it was fed to hogs in a nearby pen), they'd distill it again, raising it to 80 to 90 proof.

They also made peach and apple brandies, and there are records showing that the distillery once made persimmon brandy as well.

Upstairs from the distilling floor, there's a small museum with a few displays of antique bottles and American distilling history. Particularly helpful: you can watch a short film, "George Washington's Liquid Gold," produced by the History Channel, with a brogue-rolling actor playing James Anderson. It's worth watching for the history of distilling in early America, and also because it gives credit for so much of the heavy lifting and hot work to slave labor. Six enslaved people were known to have been assigned just to work with Anderson in the distillery.

You can buy the whiskey they're making today in the gift shop.

2   JAMES RIVER DISTILLERY

2700 Hardy St., Richmond, Va.; 804-716-5172, www.jrdistillery.com.
Hours: noon–5 P.M. Monday–Friday, 1–4 P.M. Saturday.

Away from Richmond's crowded skyline, in an industrial area near the baseball stadium for the minor-league Richmond Flying Squirrels, there's a section of warehouses called Scott's Addition. Once known for vice, particularly bootlegging and prostitution, it's now a popular bar-and-restaurant district.

James River Distillery has set up shop in a square brick building, with a tall smokestack, that originally was a steam generator.

Dwight Chew, the head distiller, was twenty-nine when I met him, a cheerful young man wearing mud boots. He offers regular tours on Saturdays and at other times by appointment, although "by appointment" can be pretty loose, he admits.

"When I'm working, people show up and I show them around. I love showing it off."

That doesn't take long: it's a single square room, two stories tall, with high windows all around. Originally started as a vodka distillery for a brand that didn't make it, the partners got the building and equipment at auction and opened in 2013.

The equipment includes a 500-gallon stainless-steel still, built by Kentucky's legendary Vendome in 1977, and a smaller "little sister," a copper Vendome built in 2004. When Chew first started, he went to Vendome's plant in Kentucky with the ID number from the larger still. An engineer took the number, went right to a drawer and pulled out the plans from thirty years ago: turns out, it was only one of two they made from that design.

Chew started as an organic farmer who got into fermenting to make his own pickles. That led him to making beer and wine, so he left farming for brewing. From there, it was a short jump to distilling.

"It's always a long, crooked road," he says.

Despite the small space, James River was making five spirits when I was there: three gins, including the only navy-strength gin in Virginia, rum, and an unusual version of aquavit.

"We are compact," he says. "But we're very efficient at what we do."

With his background in organic farming, Chew is using locally grown ingredients to come up with surprising combinations, such as cantaloupe and two kinds of hops in his American-style gin. He also works directly with local bartenders to make small-batch creations they can match to their cocktails.

"Because we're smaller, we can be flexible. We can collaborate," he says. "If Jack Daniel's wanted to make a new product, they'd have to spend two years on market research and they'd have to sell 20,000 cases." Small distilleries like James River can decide to try something, tinker with it, and just make a few cases.

"The craft beer movement really paved the way for us," he says. "It primed the consumer for different products, interesting products. That made it possible for us to make something different."

Chew worked as a line cook in college, and has always been drawn to experimenting with flavors.

"I've always loved playing with variables," he says. If you do this, do you taste that?

Case in point: his version of aquavit, the caraway-flavored liquor. One of

James River Distillery in Richmond took over an old power plant to house its stills.

the co-owners of James River is Kristi Croxton, the wife of Travis Croxton of Rappahannock Oyster Co., which is bringing farm-raised oysters to restaurants all over the South. At Rappahannock's own restaurant, they were serving oysters with aquavit on top. Chew says he thought, "That sounds weird—I'll give it a shot."

He liked the combination so much, he was inspired to come up with Oster Vit, an aquavit made with caraway, dill, fennel, and orange peel with a corn-based neutral spirit. He steeps it for forty-eight hours on the shells of Virginia river oysters that he gets from the restaurant's oyster bar. The result isn't fishy or briny, but there's a bracing mineral quality and just a hint of oyster on the back of your palate. It makes a fantastic Bloody Mary and was a runner-up in the drinks category in *Garden & Gun*'s annual Made in the South Awards in 2016.

Chew sees craft distilling as a return to a hyper-local beverage market that once thrived in America but was destroyed by Prohibition.

"Prohibition really ruined the craft world," he says. "Nobody could learn how to do it." It wasn't until craft beer started early in the twenty-first century that people started learning how to make local spirits again.

"People love seeing how it's done," Chew says. And they love the freedom that small distilleries have to play around.

"Not everybody wants to drink what their grandfathers drank."

3 RESERVOIR DISTILLERY

1800A Summit Ave., Richmond, Va.; 804-912-2621, www.reservoirdistillery.com. Tours only by appointment; tasting room 4–7 P.M. Wednesday–Friday, 3–6 P.M. Saturday.

A few blocks away from James River in the Scott's Addition district, Reservoir Distillery doesn't technically offer tours. But the lively tasting room sits in front of big windows into the distilling floor—and after a couple of cocktails, allowed at tasting rooms under Virginia's regulations, things get very friendly. Show a little interest and you're likely to get taken into the back to take a look around at the distillery, located in an old Canada Dry/Schweppes bottling plant.

All the co-owners have known each other since childhood. Today, they only make three things, and they keep them very simple: 100 percent corn bourbon, 100 percent wheat whiskey, and 100 percent rye. All the grains are grown fifty miles away at Renwood Farms.

While it's limited as far as tours, it's worth a stop to try what they have: with such focused mash bills, you can learn a lot about the tastes you'll notice in whiskeys that use a blend of grains.

See if you agree with me: their rye smells exactly like pecans.

4 CATOCTIN CREEK

120 W. Main St., Purcellville, Va.; 540-751-8404, www.catoctincreekdistilling.com. Tours are free; tastings range from $5 to $15. Hours: 1–5 P.M. Tuesday–Thursday, 1–7 P.M. Friday, noon–7 P.M. Saturday, 1–6 P.M. Sunday.

Less than two hours' drive from downtown D.C., Purcellville is a sweet antique stroller of a town right in the middle of Virginia's wine country. Catoctin (say it *cah-TOCK-ten*) is in the old Case Building, which has also been a bank, a Buick showroom, and a furniture factory.

Inside, the tasting room is large and airy, with a horseshoe-shaped bar and plenty of tables. On the weekend, you can buy light fare, like chips and dips, and watch the day-trippers go by outside the big windows.

The owners, Scott and Becky Harris, like to say that the distillery is Scott's midlife crisis. It started after they took a trip to Ireland in 2006 and he got intrigued by the making of Irish whiskey. Becky Harris has a background in chemical engineering, so she told Scott that if he came up with a business plan, she'd consider starting a distillery. He did, and Becky held up her end of the bargain, using her knowledge of chemistry to become a distiller and develop their recipes.

Today, they make a wide range of things, including brandies from several kinds of fruit. Their grape brandy, the first in Virginia since Prohibition, is aged in used French Bordeaux barrels. They also make several ryes, including a cask-strength variety and Roundstone, which is aged for two years, as well as rye-based gin and an unaged white-corn whiskey called Mosby's Spirit.

True to Virginia's style, their rye is 100 percent rye, with no corn or barley. Fermented for five days, they distill in two hybrid pot stills, "Ron Swanson" and "Barney," then use the tails from the end of the run as the base for their gin.

Their ryes have gotten wide attention, with distribution in fifteen states, including California and New York, and Australia, Singapore, Germany, and London.

"American spirits are exotic overseas," said Addie Rodgers, twenty-four, a distiller in training who gave me my tour. "Funny to think of us as exotic."

FOOD NOTE: If you need lunch after your tour and tasting, head across the street to Market Burger and bring your visit in a full circle: the beef comes from cows that eat Catoctin's spent mash.

5 MT. DEFIANCE CIDERY & DISTILLERY

207 W. Washington St., Middleburg, Va.; 540-687-8100, www.mtdefiance.com. Noon–6 P.M. Wednesday–Sunday.

You're headed into horse country now, with rolling pastures lined with white fences. In historic downtown Middleburg, owner Marc Chretien started out focusing on making local apples into hard cider, but his distilled spirits are now more than half the business.

"The long-term potential is better," he says. "Spirits are national."

Chretien's story is as interesting as what he makes: he's a retired senior

State Department military adviser who worked in war zones in the Middle East. His master distiller, Peter Ahlf, is retired from NASA. So yes, it does take a rocket scientist to make whiskey at Mt. Defiance.

Chretien and Ahlf have turned a small storefront, barely as big as an old gas station, into a busy distillery with a comfortable tasting room. There's a bottling and packaging area in the middle, and a pot still in a small side room where they make bourbon they're aging on site, corn-based unaged whiskey, vodka, gin, unaged rye, molasses-based rum—and a classic French-style absinthe.

"I'd rather do stuff that's a pain in the ass, like absinthe," Chretien says. He has a glass absinthe fountain ready to go if you're interested. He even uses handmade sugar cubes. The ingredients in his absinthe are mostly locally grown, including wormwood, except for fennel doux and anise that he imports from France.

Sadly, I missed tasting his cassis—he has tubs for soaking black currants. He's also using his cider to make apple brandy.

Most of his customers are an urban crowd from Washington who come out to spend a day in the country.

"This area is like a Virginia version of Napa," he says.

6 BLOOMERY PLANTATION

16357 Charles Town Rd., Charles Town, W.Va.; 304-725-3036, www.bloomerysweetshine.com. Noon–6 P.M. Monday–Thursday, 11 A.M.–8 P.M. Friday–Saturday. Tastings are free, but because the tasting room is small, tours are limited to groups of ten on a first-come, first-served basis. There's no tour, but the tasting is extensive and visitors can wander the grounds.

In the countryside not far from Harpers Ferry, in the far northeastern arm of West Virginia, you'll come to a gravel road leading up into the hills with a lemon-shaped sign: Bloomery Plantation.

You might hesitate to turn up the road, but press on. It curves up and up, past a hand-lettered sign urging you to use the gas pedal to make it up the final rise.

The trip is worth it, though: you'll emerge at a small farm, with lots of ground for strolling around and exploring, including a long greenhouse perched on the hill, where you can wander through the lemon trees.

Inside a small cabin that's believed to be one of the only slave cabins re-

Look for the lemon to figure out where to turn off the highway at Bloomery Plantation in West Virginia.

maining in the county, it's hard to know where to look first. Under the low, timbered ceiling, there are goofy signs ("I'd Turn Back if I Were You"), a coonskin cap dangling from a hat rack, and a polished bar crafted from wood that used to be the floor of an old bowling alley.

Bloomery Plantation is definitely not a place that takes itself too seriously.

I recounted some of the details in the introduction: how owners Tom Kiefer and Linda Losey got obsessed with limoncello on a trip to Italy, how they struggled to find anything similar in America, and how they decided to grow their own lemons and just make their own. Under West Virginia's farm-distillery rules, they have to grow at least 25 percent of what they use in their moonshine-based cordials (they actually grow about a third of their own lemons, supplementing those with lemons from California).

They've branched out from limoncello, creating a selection of ten cordials, from chocolate to raspberry to ginger.

Plenty of people seem to find their way there. When I arrived late in the afternoon during a summer visit, there was an extended family filling the tasting room on their way back from a trip to the beach, and a large and boisterous bachelor party pulled in right behind me, spilling out of several sporty cars.

In the tasting room, an energetic young Australian woman who called herself "Polly from Perth" led the tasting with a maximum number of jokes, including plenty of ideas for how to enjoy all the cordials ("Put that in your morning coffee and tell them you had drunken doughnuts for breakfast") and helpful advice ("A Big Gulp straw from 7-Eleven fits perfectly in our bottles").

Bloomery Plantation is that kind of place. Enjoy yourself. If Polly is still there, tell her thanks for the line about the doughnuts. I've used it with pleasure.

7   COPPER FOX DISTILLERY

9 River Lane, Sperryville, Va.; 540-987-8554, www.copperfox.biz.
10 A.M.–6 P.M. Monday–Thursday, 10 A.M.–8 P.M. Friday and Saturday,
1–8 P.M. Sunday. Free tours every half hour; tastings are $8.
Additional location: 901 Capital Landing Rd. in Colonial Williamsburg.

Of all the pleasant places to visit in this part of the world, a stop at Copper Fox is something special. To get there, you go up and down swooping country roads through the Shenandoah Valley and finally turn down a gravel lane to reach a cluster of wooden and brick buildings beside a rushing creek. The building that houses the distillery was built in the 1930s as an apple juice factory. In the same cluster, you'll find antique and jewelry shops, a bookstore, and a brewery, Pen Druid. The whole place, maybe a half dozen buildings, has a laid-back, counterculture feel, staffed by people who seem to love where they are and what they're doing.

Inside the distillery, there's a comfortably rumpled sitting area, like a living room shared by a bunch of college boys: well-worn leather chairs and a crazy collection of random antiques, including a sled hanging from the ceiling and a collection of stuffed foxes lurking here and there.

Copper Fox was started in 2005 by Rick Wasmund and his mother, Helen. (You'll still see the name *Wasmund's* on the trucks and on a couple of buildings in the compound.) Wasmund has a passion for using fruit wood, mostly apple and cherry, to smoke barley. Before opening Copper Fox, he traveled around Scotland, where he interned at Bowmore Distillery, one of the few distilleries in the world that smoked its own barley. Today, Copper Fox Dis-

Finding your way to Copper Fox Distillery, near the Shenandoah National Forest, will bring you to a small cluster of buildings around an old apple juice factory.

tillery uses barley, oats, wheat, corn, and rye grown by a local farmer in Northern Neck, Virginia.

While the distillery and its tasting bar are excellent, one part of Copper Fox's business makes it particularly unusual and worth your time to visit: the malting room. It's the only one like it that I've seen.

Throughout the distilling world, you hear the phrase "malted barley." Here's what it means: malting is a process of soaking barley in water just until it begins to sprout, and then drying the grain so it doesn't grow any larger. The tiny plant contains all its enzymes, which break starch into sugar for yeast to eat, so the yeast can produce (OK, excrete) alcohol.

At Copper Fox, you can go into the malting room and actually see what that means. After soaking the barley in water for two days, they spread it over one end of the room to form a blanket three to four inches deep. It looks a bit like a khaki quilt, rippled from where they draw rakes through it every couple of hours as it dries, to keep the tiny sprouts from tangling together.

After five days, they move it to the top of the kiln room, a small space next door, and spread it out over a perforated metal floor. Below, there's a roaring firebox to put out heat and a kettle grill—like a regular Weber—where they burn apple and cherry wood to make smoke, curing the barley for two days.

Copper Fox's smoked barley is in such high demand that they sell it to a long list of breweries around the country. You can see the list written in chalk on the door of the kiln room. They've opened a second facility in Williamsburg, Virginia, to keep up with the demand.

All that barley gets put to good use at the distillery, too. Wasmund started with a 100 percent barley single malt, aged in used bourbon barrels. Now they also have a rye and barley blend, aged fifteen to eighteen months, Virginia gin with a barley base, and Belle Grove, a 1790s-style whiskey—similar to the style of whiskey in George Washington's day—made with corn, oats, and unsmoked barley.

In the barrel room, whiskeys are aged in used bourbon barrels with chips of smoked apple wood added. The barrels are aged standing straight up instead of on their sides, so they can drop in the charred fruit wood and oak.

One warning about a stop at Copper Fox: the whole area suffered from notoriously spotty cell phone coverage when I was there. Make sure you download your directions for your next destination, or keep a paper road map handy. I learned that the hard way. Then again, Sperryville is such a pretty mountain town, you might prefer an excuse to get stranded there for a year or two.

8   SMOOTH AMBLER SPIRITS

745 Industrial Park Rd., Maxwelton, W.Va.; 304-497-3123, www.smoothambler.com. Free tours and tastings; tastings are 2–5 P.M. Monday–Friday, 11 A.M.–3 P.M. Saturday. Tours are at 2 and 4 P.M. Fridays and noon and 2 P.M. Saturdays.

One night after a dinner at the James Beard House in New York, I ended up down in the kitchen with the celebrating chefs, who were passing around a bottle of whiskey to mark the occasion. The bottle was the chef's choice: Smooth Ambler's Old Scout, a whiskey that's gotten very popular in the food world.

Smooth Ambler is a thriving distillery on a back road in Maxwelton, just a few miles from the town of Lewisburg, voted "The Coolest Town in America" by the website Thrillist. The distillery has gained a reputation, using all-local corn, wheat, and rye, most of it grown in Monroe and Greenbrier Counties.

There's something else that makes Smooth Ambler notable: it was one of the first distilleries to be open about starting out with whiskey made at the large MGP distilling facility in Indiana. While some distilleries used to keep that quiet, Smooth Ambler was always honest about it, calling those products "merchant bottle products." (For more on that, see the sidebar "MGP: Right or Wrong" in Liquor Trail 5.)

Smooth Ambler is now moving away from outsourced whiskey, producing enough to create its whiskeys entirely on its premises and filling several large black-and-red warehouses that now dot the property, each building positioned to get the best conditions in summer and winter.

When you visit, you'll find an elegant tasting room and gift shop, and a tour through a distilling room that is at full production from a huge, custommade Vendome continuous still.

TRAVELER'S NOTE: If you stop overnight in this area, go into Lewisburg and find the General Lewis Inn, which dates to the 1820s. The rooms are a little creaky, but the high beds are very comfortable, and the whole place is furnished with antiques. The owners also operate an excellent small restaurant up the street, the Stardust Cafe.

Other Craft Distilleries in Virginia and West Virginia

VIRGINIA

A. Smith Bowman. 1 Bowman Drive, Fredericksburg, Va.; 540-373-4555, www.asmithbowman.com. Free tours 10 A.M.–4 P.M. Monday–Saturday.

Belle Isle Craft Spirits. 615 Maury St., Richmond, Va.; 804-404-9550, www.belleislecraftspirits.com. Moonshine distillery with several flavors. Tasting room and tours not yet available.

Belmont Farm Distillery. 13490 Cedar Run Rd., Culpeper, Va.; 540-825-3207, www.belmontfarmdistillery.com. Farm-based unaged whiskey and moonshine distillery. Tours available 11 A.M.–4 P.M. Tuesday–Saturday; closed Sunday and Monday.

Bondurant Brothers Distillery. 9 E. Third St., Chase City, Va.; 434-738-7372, www.bondurantbrothersdistillery.com. Moonshine distillery with a tasting room still being set up.

Chesapeake Bay Distillery. 437 Virginia Beach Blvd., Virginia Beach, Va.; 757-498-4210, chesapeakebaydistillery.myshopify.com. Tours noon–4 P.M. Saturdays for free; tastings $7. Includes vodkas, lemon liqueur, and rum.

E. Wright and C. Wallace Distilleries. 2209 Gladesboro Road, Hillsville, Va.; 276-398-2257, www.boarcreekwhiskey.com. Makers of Boar Creek whiskey. Call for tour details.

Ironclad Distillery. 124 Twenty-Third St., Newport News, Va.; 757-245-1996, www.ironcladdistillery.com. The tasting room was still under construction when we checked, so they aren't offering regular tours yet. But they will do them by appointment.

River Hill Distillery. 356 Ruffners Ferry Rd., Luray, Va.; 540-843-0890, riverhilldistillery.com. Tours 11 A.M.–5 P.M. Saturdays; other times by appointment.

Silverback Distillery. 9374 Rockfish Valley Hwy., Afton, Va.; 540-456-7070, www.sbdistillery.com. Vodka, gin, and grain spirits, with rye and bourbon on the way. Open only in summer; noon–5 P.M. Monday and Thursday, noon–6 P.M. Friday and Sunday, noon–7 P.M. Saturday; closed Tuesday and Wednesday.

Virginia Distillery Co. 299 Eades Ln., Lovingston, Va.; 434-285-2900, www.vadistillery.com. 11 A.M.–6 P.M. Monday–Saturday, noon–6 P.M. Sunday. Tours range from $12 to $32 depending on tastings.

Vitae Spirits. 715 Henry Ave., Charlottesville, Va.; 434-270-0317, www.vitae spirits.com. White and gold rums and gin. Tours 2–9 P.M. Wednesday–Friday, 10 A.M.–9 P.M. Saturday, noon–6 P.M. Sunday.

Williamsburg Distillery. 7218 Merrimac Trl., Williamsburg, Va.; 757-378-2456, www.williamsburg-distillery.com. Microdistillery focusing on eighteenth-century methods, with gin, rum, and bourbon. Free tours and tastings every thirty minutes from 1:30 P.M.–7:00 P.M. Wednesday, Friday, and Saturday.

WEST VIRGINIA

Appalachian Distillery. 3875 Cedar Lakes Dr., Ripley, W.Va.; 304-372-7000, appalachian-moonshine.com. Flavored moonshines in a variety of flavors. Includes a tasting room. 8 A.M.–6 P.M. Monday–Thursday, 8 A.M.–8 P.M. Friday–Saturday; closed Sunday.

Pinchgut Hollow Distillery. 1602 Tulip Ln., Fairmont, W.Va.; 304-366-9463, www.hestonfarm.com. On the grounds of Heston Farm Winery.

Moonshines and whiskeys in unusual bottles, shaped like pigs standing on their heads. 11 A.M.–9 P.M. Monday–Saturday.

Recipes

SMOKED MINT JULEP

I've spotted Copper Fox products at craft-cocktail bars all over the South. One night in Charlotte, soon after the opening of the restaurant Haberdish in the North Davidson Street arts district, I was delighted to spy a Smoked Mint Julep on the drink menu. Instead of using a smoke infuser, as many bars do, Colleen Hughes, the beverage manager, was using Copper Fox Smoked Rye to put a sophisticated spin on one of my favorite drinks.

MAKES 1 DRINK

2 ounces Copper Fox Smoked Rye
1/2 ounce Branca Menta (see note)
3/4 ounce mint-infused simple syrup (recipe below)
2 sprigs of fresh mint

Place one mint sprig in the bottom of a metal julep cup and muddle (crush) gently. Top with some crushed ice. Slowly add the rye, Branca Menta, and syrup. Stir gently, trying not to lift the mint leaves off the bottom. Top with more crushed ice and garnish with the second mint sprig.

Mint Simple Syrup

Combine 1 cup sugar and 1 cup water in a small saucepan. Bring to a simmer and stir until the sugar is dissolved and the syrup is clear. Remove from heat and stir in 2 tightly packed cups of mint leaves. Let stand for 30 to 45 minutes, or until cool. Strain the syrup and discard the mint leaves. Refrigerate the syrup for up to six weeks. (From *Bourbon: A Savor the South Cookbook*, by Kathleen Purvis.)

NOTE: Branca Menta, made by the Italian amaro maker Fernet-Branca, is an herbal liqueur with mint essential oil. It's available in many liquor stores that carry Italian amari.

LIMONCELLO MARGARITA

From Bloomery Plantation in Charles Town, W.Va.

MAKES 1 DRINK

1 ½ ounces fresh lime juice
1 ½ ounces limoncello
2 ounces 100% agave tequila
Rimming salt

Fill a cocktail shaker with ice and add the lime juice, limoncello, and tequila. Shake well. Wet the rim of a cocktail or margarita glass and dip in rimming salt. Strain the cocktail shaker into the glass and serve cold.

HEDGEMAN'S REVIVER

One of the most unusual spirits you'll encounter is the Oster Vit, a variation of aquavit made by James River Distillery in Richmond. When I asked co-owner Kristi Croxton how to use it in cocktails, she suggested it makes a good substitute for vodka in things like a Southside, gimlet, or Moscow mule. This recipe, a twist on a Corpse Reviver, is from Paul Kirk at the restaurant Rappahannock, owned by Croxton's husband, Travis.

MAKES 1 DRINK

1 ½ ounces Oster Vit
½ ounce fresh lemon juice
½ ounce Cointreau
½ ounce Cocchi Americano
Absinthe rinse
Lemon twist and fresh dill sprig (optional; garnish)

Fill a cocktail shaker with ice and add the Oster Vit, lemon juice, Cointreau, and Cocchi Americano. Shake vigorously.

Pour a little absinthe in a chilled coupe or cocktail glass, swirl it around to coat the glass, and pour out the excess.

Strain the contents of the shaker into the glass. Garnish with a lemon twist and a sprig of fresh dill.

How to Taste:
No Swirling, No Spitting

Tasting liquors is the best part of touring distilleries. But it's not exactly the same as tasting wine at a winery. A few things to know:

Sizes. Most states set limits on the amount of alcohol you can taste on a tour. It ranges widely, but usually you get quarter-ounce pours of four or five products, totaling no more than two to three ounces. Some distilleries are very free-handed with their pours; others adhere strictly to the rules, limiting the number of things you can try at a distillery that might make eight or ten products.

No spit buckets. There may be a bucket for dumping liquor you don't want to drink if you're driving (or just don't like it), but you need to swallow, at least a little, to fully evaluate the flavor and aftertaste.

Glassware. The containers you'll be given your tastes in range from the traditional Glencairn glass, shaped sort of like a miniature hurricane glass, to shot glasses or disposable plastic cups. If it's a shot glass with a logo, you usually get to keep it as a souvenir. If you notice small, clear plastic cups that look like communion cups, they might be; more than one distiller told me the first company to call when they opened sold church supplies.

Smell first. I see this over and over at tasting bars. People get their sample and immediately start to swirl it vigorously, the way they were taught in wine tastings. But liquor isn't the same as wine. Instead of agitating to re-

At a tasting bar, you'll get your samples in everything from plastic cups to small glencairn glasses, which help to capture aromas and direct them up to your nose.

lease the esters, or flavor compounds, you want to contain them so you can fully evaluate them. I like to cover my glass for a few seconds, then uncover it. Next, put your nose well into the glass and open your mouth slightly while you breath in. (I've heard of a connoisseur who would close one nostril at a time while smelling. You don't need to go that far.) Take one sniff and try to pick out one flavor (vanilla, caramel, smoke, honey, orange, chocolate), then take a second sniff and try to pick out another.

Sip. Don't rush, and don't throw it back. Try to move it all over your mouth before you swallow, even swishing (just a little), to discover what you notice. Pause for a minute after you swallow and think about whether it has an aftertaste that lingers, or a bite that hits the back of your throat.

Trust yourself. Guides at distillery tasting bars are prone to telling you what you'll taste as they pour ("you'll taste caramel, vanilla, and a hint of the owner's dog"). That can be misleading. Try to ignore the power of suggestion and focus on what you smell and taste. You're the one deciding what you want to buy.

North Carolina

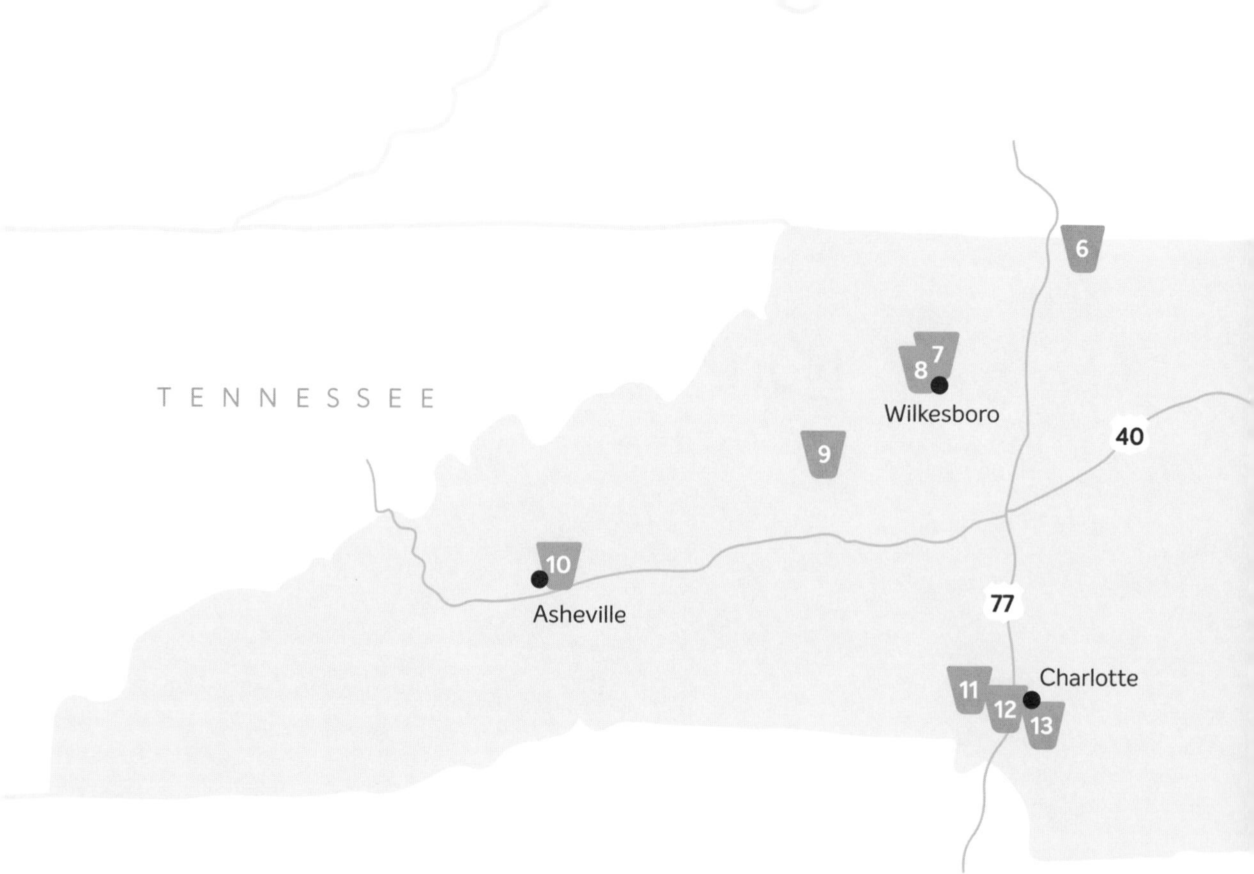

TENNESSEE

Wilkesboro

40

Asheville

77

Charlotte

6

7

8

9

10

11

12

13

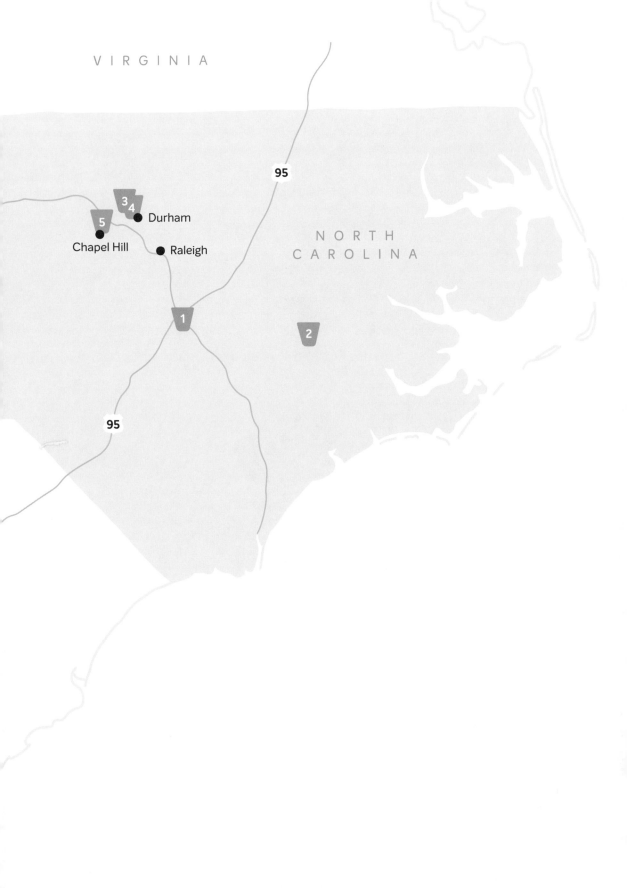

VIRGINIA

95

NORTH
CAROLINA

3
4 ● Durham

5
● Chapel Hill

● Raleigh

1

2

95

FROM A REGION THAT was almost synonymous with illegal liquor until well into the twentieth century, what has happened in North Carolina in just a few years has blown the lid off the moonshine jar.

For decades, any form of alcohol was a loaded issue there. Much of North Carolina's attitude and history was shaped by a strong religious culture on the one hand and a proud, defiant culture of moonshining on the other.

In the mountains that ripple the western side of the state and the isolated farm communities that dot the east, there was a constant struggle between the two points of view right through the twentieth century—and plenty of overlap as well. An awful lot of people were regulars both at the local boot-legger's on Saturday and in church pews on Sunday.

The illegal way of life was so entrenched that the state still regulates alcohol sales with a tight grip, with most of North Carolina's liquor sales only allowed through a state-controlled Alcoholic Beverage Control store system.

Still, control never worked completely. In the North Carolina mountains, running illegal liquor past law enforcement became such a part of life that drivers started holding local races with their souped-up cars, giving rise to stock-car racing, the roots of what we now know as NASCAR.

Illegal liquor wasn't confined to the mountains, either. Two counties in North Carolina, Wilkes County in the northwest and Johnston County in the east, still debate which one holds the right to be called "the capital of moonshine." Wilson County, where tobacco auctions once drew buyers from all over the world, lays claim to the title as well. You encounter people who are cautiously open about illegal creation that still goes on today.

In 2005, though, something happened in North Carolina that is still reshaping state liquor regulations. Pop the Cap was a grassroots movement that got North Carolina's laws changed to allow the sale of beer with up to 10 percent alcohol by volume.

At the time, beer fans expected an increase in higher-alcohol imported beers. Instead, what happened surprised everyone: the craft brewing move-

Brian Call of Call Family Distillers in Wilkesboro, N.C., is the seventh generation in his family to make moonshine.

ment, which had been strongly focused in Colorado, California, and Washington State, suddenly took off in North Carolina. Within a few years, hundreds of microbreweries opened from Raleigh to Asheville. In Charlotte alone, there were twenty-seven in 2016, with twenty-one more expected by the end of 2017.

The interest—and the potential for tax revenue—didn't escape the attention of the state government or people who wanted to make liquor. By 2010, distilleries started to open as well and new laws are helping them grow. North Carolina now allows distilleries to have tasting rooms. After starting with a limit of one bottle per person per year, the state raised the tasting-room limit to five bottles in 2017. That has helped distilleries to create a following that drives sales to the state-controlled stores. (Be prepared to leave your driver's license number if you're going to buy. That's how distilleries ensure that you wait a year before you buy more from them.)

When you visit, your business helps the distilleries survive: under state regulations, the ABC can remove a product from its shelves if it doesn't meet a minimum yearly amount of sales. Without sales in the state system, a distillery can't make enough money on the amount it can sell to visitors. At many tasting rooms, you'll see little displays directing you to the nearest ABC store that carries its products, in the hope you'll like what you tried enough to make a special stop to buy more.

In response, distilleries that never planned on retail sales have gotten busy building tasting rooms and adding tours to take advantage of it.

There is still a struggle between the wholesale distributors and the makers of both beer and liquor, but there's a reason this is the longest trail in the book. There's so much to find and such a variety of things being made that North Carolina is challenging Virginia and Kentucky as the center of southern craft distilling.

For this trail, we'll start in eastern North Carolina, head across the state toward the mountains, and then circle down to Charlotte.

1 BROADSLAB DISTILLING

4834 N.C. Hwy. 50 South, Benson; 919-207-1366, www.broadslabdistillery.com. Tours: Noon, 2 P.M., and 4 P.M. Thursdays, Fridays, and Saturdays. Tasting room and gift shop open noon to 5 P.M. Thursday–Saturday. Tour: $12 and includes a shot glass. (The rate is lower for groups of 10 or more.)

Let's start with a true farm distilling experience. At Broadslab, about forty miles south of Raleigh, your tour starts at a cabin built on the site of an old farm stand. Then you climb aboard a comfortable trailer outfitted with benches—sort of like a front porch on wheels—and get pulled by a tractor for the trip through the cornfield, into the woods, and finally to the distillery.

Jeremy Norris, who's in his early forties, grew up on this land, raised by his grandfather. It was a vegetable farm in those days, with a produce stand sheltered by an oak tree by the highway, where they did a steady business selling corn, tomatoes, and watermelons to people who passed by on the way to the beach.

The Norris farm was originally 400 acres, but like a lot of farms, they sold off chunks to stay afloat, eventually dwindling to just the little area by the road: "Cut up, sawed off, cut up, sawed off," he says.

On those afternoons at the farm stand, Norris's grandfather used to tell him stories about the moonshining that used to go on. His great-grandfather made whiskey and hauled it to the beach to sell, and his grandfather had made plenty himself.

"He told me the same stories—over and over and over. He always told me he had a doctorate degree in moonshining." His grandfather used to tell Jeremy he should open a moonshine museum on their land. Jeremy just laughed.

Still, his grandfather was adamant about keeping the farm alive, and Jeremy wanted to try. In 2005, he bought back sixty acres to farm. Financially, he discovered, it just didn't work.

"You can't farm small anymore," he says. "Unless you've got a niche product."

With legal moonshine taking off, his grandfather's stories gave him an idea about what that niche product could be: Norris had land to grow his own ingredients—and his grandfather to teach him how to make it.

It turned out to be harder than he thought, of course. It took four years to get up and running. He built the still based on a drawing his grandfather made.

It was 2012 before he had something he could sell. His grandfather died thirty days after they opened. Norris pushed on, turning his grandfather's house by the highway into a pretty tasting room, with a shady front porch for visitors and a gift shop and tasting bar. Eventually, he hopes to add a restaurant.

For the tour, you climb onto the wagon, he revs up the tractor and you're hauled through fields of old-variety white corn, then onto a dirt road. In the 1700s, the road was paved with broad slabs—trees sawed in half from a local sawmill—to keep wagons from miring in mud. (Keep your eyes peeled: Jeremy set up displays along the way of the old days of moonshining.)

The distillery itself is a small building decorated with old farm implements, tobacco baskets, and burlap sacks. The pot still and doubler, based on his grandfather's drawings, are classic designs. The water comes from his well, unusual in this part of the world, where water quality can be poor. It turns out the land is on a sand bed, which filters the water and makes it very clear and clean. On the land on either side of Norris, the water isn't as good. Maybe there was a reason all those Norris ancestors had such success making moonshine.

His lineup includes Hanging Dog, a 90-proof straight corn whiskey he makes for Wayne Nix of the TV show *Moonshiners*; Legacy Shine, a sweeter, smoother whiskey made from corn and malted barley, two rums, a barrel-aged whiskey; and Legacy Appleshine made with juice from organically grown Vermont apples. He's also aging a bourbon made from Indian corn.

Norris does everything himself, even charring his own staves to line his barrels.

"We are low-tech," he says. "That's the way I want to keep it."

2   MOTHER EARTH SPIRITS

311 N. Herritage St., Kinston; 252-208-2437, www.motherearthspirits.com.
Free tours hourly from 10 A.M.–5 P.M. Tuesday–Friday and 1–8 P.M. Saturday.

With so much of the state's alcohol production focused on beer, it's inevitable that the two worlds will start to overlap. Mother Earth Brewing, in a large brick building in downtown Kinston, has quickly attracted a following, and now it has branched out, with a small distillery in a side room.

Thanks to Vivian Howard's restaurant, Chef & The Farmer, and her PBS show, *A Chef's Life*, food-focused visitors have been flocking to Kinston's tiny downtown. If you knew Kinston before Howard and her husband, Ben Knight, opened their restaurant, you would never have guessed that it would become a tourist destination. Just a few years ago, it was practically deserted, typical of so many small eastern North Carolina towns that dwindled when the tobacco industry disappeared.

Mother Earth's building is just a block from Howard's food business, in a brick building that was started in the 1800s as a mule stable.

Inside, the bar is cool and contemporary, decorated in a midcentury modern style and usually packed on a Saturday afternoon with beer drinkers grabbing a brew and waiting on a brewery tour. But you also can sign up for the distillery. Owners Stephen Hill and his son-in-law, Trent Mooring, added it in 2014 to broaden their business model. It's believed to be the only solar-powered distillery in the country.

Brewing and distilling are compatible activities. After all, when the mash of grains for whiskeys is fermenting, the low-alcohol mixture is sometimes called distiller's beer, so the first few steps are similar. But under state rules, you can't use the same equipment for brewing and distilling, and you can't do it in the same room. So the distillery is tucked into a single room, near the gift shop, on the other side of the large brewing room.

They buy neutral-grain spirits for their gin and then age it for a week in small barrels with orange and lemon peel, Szechuan peppercorns, grains of paradise, coriander, and orris root. They also make silver rum from black-strap molasses and sugarcane, and they're aging a whiskey, using 100 percent malted barley and aiming for an Irish poteen style.

The distillery tour doesn't take long, and then you can sign up for the brewery tour as well. The two can fill an afternoon in Kinston while you're waiting for a crack at a table at Chef & the Farmer or a dozen oysters at Boiler Room, Howard's second restaurant, located in a nearby alley.

"If tourists aren't here for Mother Earth, they're here for the Chef & the Farmer," said my tour guide, R. J. Smith, twenty-one, a native of Kinston.

"If they're not here for the Chef & the Farmer, they're here for Mother Earth."

3 DURHAM DISTILLERY

711 Washington St., Durham; 919-937-2121, durhamdistillery.com. Tours: 4 and 6 P.M. Saturdays and most Friday evenings as well (if you call, you also can often get a morning tour on Saturday by request); $15 with a tasting and souvenir shot glass. (They also offer $10 "quick tours" on Fridays from 6–9 P.M. and Saturdays from 3–7 P.M. that only include a guided tasting, not a walk through the distillery.)

No one is suggesting you should encourage a child to develop a taste for gin. But Melissa and Lee Katrincic have Melissa's grandfather to thank for their line of gins called Conniption.

Melissa spent a lot of time with her grandparents when she was young, and her grandfather used to give her the olives from his cocktail-hour martini, with a little gin flavor still clinging to them. By the time she was grown, she was a passionate fan of gin. (When she was rambunctious, her grandparents used to tell her "not to throw a conniption," so they gave her the name for her gins, too.)

Katrincic's husband, Lee, is Croatian by heritage and a pharmaceutical chemist by training, so he knew the science of distilling. When she was facing a layoff from her marketing job in the economic downturn, they started looking for something they could do together that played to both their strengths, her knowledge of marketing and his laboratory training.

They were making a long drive back from Florida when it hit her: "Why don't we make gin?" Before they even got back to Durham, she was on her smart phone in the car, lining up a name and figuring out how to apply for a license.

They built a small distillery in the Central Park district of Durham, just up the block from the old Durham Athletic Park, the original home of the Durham Bulls, made famous in the movie *Bull Durham*. Still a little industrial, the neighborhood also is becoming a beverage destination: Fullsteam Brewery is nearby, and there's a small bar next door, Blue Note Grill, which uses Conniption in its cocktails.

The Katrincics are serious about gin, making a range of styles, including a Navy Strength, with plans to add an Old Tom sweetened gin and a barrel-rested version as well. They're not making their own neutral grain spirits, they're using an ethanol to get a completely flavorless base to add their own botanicals using vapor infusion, similar to Hendricks and Sapphire.

Lee also uses an eye-catching glass sphere, called a rotary evaporator, to make cucumber, fig, and honeysuckle vodkas, and they have a line of coffee, chocolate, and mocha liqueurs.

The tour includes a useful tasting exercise. They put out bottles of botanicals, including dried honeysuckle, juniper, coriander, and orris and angelica roots, so you can smell them separately and then sip samples to help you learn to isolate the flavors.

"We thought the technology of gin needed to be pushed," says Melissa. "We wouldn't just make gin to make gin. We are passionate about gin."

4   THE BROTHERS VILGALYS SPIRITS

803 Ramseur St., Durham; 919-617-1746, www.brothersvilgalys.com. Tours: 6 P.M. Thursday–Friday and hourly 2–5 P.M. Saturday. Standard tour and tasting is $5 including a shot glass. Cocktail classes also are available; check the website for times.

In a small warehouse on the other side of downtown Durham from Durham Distillery, Rim Vilgalys cheerfully calls his business "a fake distillery." Still, what he's making is so unusual, it's worth a stop to experience it.

Vilgalys is from a Lithuanian family, and his father's tradition was making krupnikas, a traditional honey-based liqueur made in Polish and Lithuanian homes. Rim learned to make it watching his dad, and he kept it up as a hobby when he went off to college. "It got me invited to a lot of parties," he says.

After school, he moved back to Chapel Hill and decided to try making krupnikas as a business, using a Kickstarter campaign to raise money and taking business classes at a local technical school. (His brother was originally going to partner with him but ended up moving to New Zealand at the last minute, so there's really only one brother in the Brothers Vilgalys.)

Using local honey from Person County, Vilgalys started making a better version of krupnikas than his dad made with Everclear. He buys ethanol and uses the base of a still with electric burners to heat the alcohol and honey.

"It comes down to honey, booze, and spices," he says.

Rim, an Ultimate Frisbee fanatic who looks a lot younger than he actually is, is also a smart businessman. Since krupnikas is usually sipped in the winter, he's expanded his business to summer by adding a line of liqueurs in a crazy range of flavors with crazy names: Zaphod, Beebop, Beatnik, and Jabberwock, flavored with beets, star fruit, passion fruit, chicory, and even chili pepper.

Krupnikas is now available in Georgia, South Carolina, Delaware, Maryland, and Chicago. Vilgalys finds a lot of customers in cities like Chicago with large Eastern European populations, but he's also finding fans among mixologists at craft-cocktail bars. I've spotted it at bars all over the South, often used in cocktails with ginger or coffee, or to add a honey spice flavor to drinks with rum, whiskey, and tequila.

Krupnikas has come a long way, he says; at his dad's parties, all they ever did was pour a glass and say, "Here, mix that with your face."

5   TOP OF THE HILL DISTILLERY

505c W. Franklin St., Chapel Hill; 919-699-8703, www.topodistillery.com.
Tours: 7 P.M. Fridays and at varying times on Saturdays. Tours are $20.
(Note: the distillery faces Franklin Street, but the parking lot and entrance
are around the corner on South Graham Street.)

The story of TOPO Organic Spirits started around the corner from the distillery in the old Chapel Hill News building at Franklin and Graham streets, at a restaurant and brewery called Top of the Hill.

Owner Scott Maitland served in the Army, and then came to the University of North Carolina for law school, planning to settle in Chapel Hill, where school traditions are sacred and every heart beats Carolina blue. In 1996, he heard that a TGI Fridays was going to open at Franklin and Columbia streets, the Times Square of Chapel Hill. Maitland was so appalled, he grabbed the location himself and opened a restaurant, Top of the Hill, with a balcony overlooking Franklin Street.

Maitland eventually added a brewery, and became the second brewery in the state to sell microbrews in cans. He was planning to build a second brewing facility "as a lark," but then noticed that distilling was starting to take off.

A few blocks from the restaurant, he found the old newspaper building, which had tanks for holding ink and an industrial flammability rating that made it ideal for a distillery with a tasting room and events space.

To make his alcohol, Maitland wanted to use only grains grown organically in North Carolina. Part of Maitland's interest in distilling was in the idea of boosting the local economy by keeping the money from local agriculture in town.

"Our money is not 'sticky' in North Carolina," he says. Too much of what's grown goes to other states to be made into products. Instead, Maitland is passionate about using local agriculture to make local liquors. He uses things like the state's soft red winter wheat, which is great for distilling because it's low in protein, and his Piedmont Gin is flavored with all North Carolina-grown botanicals. He also makes a wheat vodka, an all-wheat whiskey, and 8 Oak, a whiskey flavored with new wood chips from American and French oak using different levels of charring.

"I think we're entering the first golden age of distilling," he says, part of the bigger craft-food movement that's getting away from mass-produced products dominated by a few large companies.

"Hey, there's great stuff [already being made by mainstream American distilleries], I agree. But it's all the same stuff."

6 MAYBERRY SPIRITS

461 N. South St., Mount Airy; 336-719-6860, www.mayberryspirits.com.
Tours: noon–5 P.M. hourly Fridays and noon–5:30 P.M. every half
hour Saturdays. $10 with souvenir shot glass.

If you're a fan of *The Andy Griffith Show*, you may already know about Mount Airy. Actor Andy Griffith was born there and used it as the inspiration for his fictional creation, Mayberry. Today, Mount Airy has turned nostalgia around the show into its primary business, with pork-chop sandwiches at the Snappy Grill and, yes, a Floyd's Barber Shop.

You don't have to whistle the theme song to enjoy a stop, though. It's also the location of a massive granite quarry, and the streets are lined with gorgeous Victorian-era houses with ornate granite foundations. Between that and a hunt for sonker, a local version of cobbler, it's a nice place to spend a Saturday afternoon.

Mayberry Spirits is just as quirky as the town itself. Owner Vann McCoy has one of the stranger back stories you'll come across. Originally a student of astrophysics, he became a Cistercian monk instead and spent twenty-five years in monasteries in Switzerland and Ireland. He left the order a few years ago and came back to Mount Airy to take care of his aging mother, and ended up starting a distillery specializing in what he calls "white whiskey," based on sorghum, the heirloom grain that is being rediscovered in the South.

Located in a rustic building made from reclaimed wood, the distillery includes a well-stocked gift shop with local crafts where you can browse while you wait to join one of the popular tours. (If you're a baker, take a look at their line of cake flavorings they make themselves.)

McCoy and his crew like to ham it up for visitors—he sometimes dons a long, fake "mountaineer" beard for his talk on distilling basics. But true to his science background, McCoy and his other tour guides all do an excellent job of explaining the process, including the breakdowns of temperature points for heads, hearts, and tails. Aunt Bee would proud of how well they've done their homework.

7 COPPER BARREL DISTILLERY

508 Main St., North Wilkesboro; 336-262-6500, www.copperbarrel.com.
Tasting room and gift shop hours: 10 A.M.–7 P.M. Monday–Saturday, 1–5 P.M.
Sunday; tours available every half hour; $10 ($44.50 for a personal
tour by distiller Buck Nance; two-person minimum).

North Wilkesboro, in the hilly Piedmont region just east of the North Carolina mountains, isn't a place where "hipster" comes to mind. It's the last place you'd expect to find gourmet moonshine.

Like a lot of small towns, the end of furniture manufacturing left empty, brick buildings in North Wilkesboro's downtown. So Copper Barrel is a surprise when you pull up. Built in an old furniture factory, the front room is a slick gift shop with a tasting bar crafted from wood from the Old Crow Distillery in Kentucky and fronted with railroad timbers from the site of a train

Copper Barrel Distillery is built in what used to be a furniture factory in downtown North Wilkesboro, N.C.

wreck in Danville, Virginia, that was immortalized in Johnny Cash's song "The Wreck of the Old 97."

North Wilkesboro isn't at all where George Smith, a native of Vermont who still lives in Charlotte, expected to end up running a moonshine distillery. In his forties with a shaved head, a neat beard, and an earnest demeanor that wouldn't be out of place behind a bar in Brooklyn, Smith started out as a project manager for IBM. At heart, though, he was a food fan who loved agriculture and had a dream of making bourbon.

At a food event in Charlotte, he met Bill Samuels, the retired owner of Maker's Mark, who was impressed enough by Smith's seriousness that he told him, "If you ever decide to do it, come see me and I'll help you make the connections."

With that encouragement, Smith decided to do it. He went to Kentucky, where Samuels arranged visits to barrel cooperages and Vendome, the famous still maker.

Smith couldn't afford to build a distillery in Charlotte, where real estate costs are high. Like the project manager he was, he got methodical: he made a spreadsheet with fifteen criteria, from local demographics to water sources. He narrowed his list to twelve towns, with Asheville and Wilson at the top. Other towns started to court him, though, including North Wilkesboro, long known as a center of illegal moonshine. Smith wasn't interested in moonshine, but a local official said to him, "Look, you're going to need a master distiller. Let's go out and meet a guy."

Now, Smith knew one thing for sure: he wanted to make bourbon, not moonshine. He hated moonshine, he says. Every one he had tried tasted like kerosene. But he agreed to go out to the country and meet Buck Nance, a seventy-year-old mechanic who built stills and made a little 'shine himself on the side.

Nance asked Smith why he was so set on bourbon. Because I hate moonshine, Smith said. Nance told him to wait a minute, walked off into the woods and came back with a quart jar of moonshine. It was the best Smith had ever tasted. Nance told him to wait again, walked into the woods and came back with a second quart jar. It was better than the first.

Smith was hooked. Nance showed him the plans for a steam-injection distillery based on a drawing his father had brought back from World War II. At first, the plan was for Nance to build the still and train Smith to run it. But Nance got hooked on Smith's dream, too. The two men became an unlikely partnership, the older man in charge of making the liquor and the younger man handling the regulations, marketing, and distribution.

"He gets to focus on his passion, making liquor, and leave the paperwork and paying taxes to me," Smith says.

The moonshine Nance makes isn't like any other moonshine I've encountered. It's much smoother and even works in cocktails, where moonshine's distinctive flavor is often too strong to blend well. Nance and Smith also have come up with fruit versions based on traditional "bounces" people used to make, adding whole fruit like cherries, strawberries, and blueberries.

"The biggest hurdle we've had to overcome is convincing people we have great moonshine," Smith says. So many places make moonshine as a novelty that it's gotten a bad reputation.

Without being hokey about it, Smith loves to showcase Nance as a true craftsman of Wilkes County. You can pay extra for a personal tour with Nance himself, overalls and all.

"I'm passionate about keeping moonshine history in North Carolina alive," Smith says. "This is truly traditional Wilkes County moonshine."

1611 Industrial Dr., Wilkesboro; 336-990-0708, www.callfamilydistillers.com. Tours: 8:30 A.M.–5:30 P.M. Monday–Friday, 9 A.M.–3 P.M. Saturday. $10 for a full tour and tasting, $5 for a half tour and tasting (all tours come with a free soft drink and a Moon Pie).

Yes, there is a North Wilkesboro and a Wilkesboro in Wilkes County. Separated by the Wilkes River, the two towns have separate downtowns, separate town halls, and even separate police departments.

What connects them is moonshine history. Any visit to Wilkes County is a deep dive into moonshine, both tasting it and learning about the colorful history of it.

In an industrial area of Wilkesboro, surrounded by auto-body shops and metal-siding factories, Brian Call has built a distillery that he's quick to admit is the continuation of the work of generations of Calls. It's just that the current business is no longer likely to land them in prison.

Call is the seventh generation of his family to make liquor. The first was the Reverend Daniel Call, who moved to Lynchburg, Tennessee, from Wilkes County to run a church and store, taking his still along with him. The congregation objected, so he sold his still, and the slave who ran it, to a young man named Daniels, whose family called him Jack. Brian's own son is now majoring in business at Western Carolina University so he can run the distillery someday—generation eight.

"Call is a big name in the moonshine business," Call says. "Times changed. We thought it was time to go legal. This is the way my dad made it, my granddad made it."

Call's father was the late Willie Clay Call, a legendary moonshine maker and runner who was dubbed "The Uncatchable" by local federal agents and "The Bull of the Woods" by locals. He had gotten out of the business (well, mostly) by the time Brian was born, but the family history is all over the distillery: in the front, at the check-in desk for tours and in the tasting room, there are family photos (including mug shots), and a series of paintings featuring guys like Willie Clay Call and stock-car driver Junior Johnson hanging around campfires by stills and handing off mason jars behind barns.

The tour includes a good explanation of the process, including a custom still built by Vendome. Call also uses the steam-injection method, similar to the one that Buck Nance uses at Copper Barrel.

"That's a Wilkes County thing," Call says. The wood fires that heated pot stills in the woods were hard to regulate and could scorch the liquor and ruin the still, or even cause explosions. Steam is easier to control, and there is evidence of steam-injection stills that date to the nineteenth century.

Besides the distilling operation, which uses white cornmeal ("Dad said white corn for whiskey and cornbread, yellow for slopping the hogs") from Linney's Mill, a local water-run gristmill, the tour also showcases a lot of historic objects, including two fine cars. One is a black 1940 coupe with extra carburetors, a souped-up engine, and extra stiff springs so it wouldn't sag when it was loaded—a dead giveaway for law enforcement. The real beauty, though, is a baby-blue 1961 Chrysler New Yorker, impeccably preserved, with Willie Call's sunglasses still in a case on the dashboard. It was his favorite "whiskey" car. He'd load up, put on a suit and sunglasses, and drive through town in broad daylight to make his moonshine deliveries.

"They'd think he was a judge or a lawyer. He used to haul two or three loads a night."

9 CAROLINA DISTILLERY

1001 West Ave. NW, Lenoir; 828-499-3095, www.carolinadistillery.com.
Tours: on the hour noon–5 P.M. Friday and Saturday; $10 including samples
and a souvenir shot glass. Other tour times available by appointment.

On the edge of the mountains, about sixteen miles from Boone, Lenoir is trying to bring back its historic district, helped by day-trippers who stop on drives to view fall leaves and visitors who come from around the globe in September for what the town claims is the world's largest tattoo festival.

In 2006, Keith Nordan and his partners opened Carolina Distillery to make Carriage House apple brandy. After outgrowing a small building that actually was the site of a carriage maker, he's relocated to a 40,000-square-foot building that opened in the 1930s as a furniture store. The front room is big and open, with a bandstand and lots of space that he rents for receptions.

Nordan worked for Lowe's for more than thirty years, specializing in custom log homes. He built the wooden bar in the tasting room himself—"not a nail or screw anywhere in it."

When the market for custom log homes dried up, Nordan was looking for something he could create and market. He grew up in Fayetteville and knew nothing about the mountain history of making liquor, but a friend, tattoo artist Chris Hollifield, grew up around illegal liquor and was making wine as a hobby. Nordan tried some of the wine—"it was terrible," he says—and asked Hollifield, "What else can you make?"

"We chose apple brandy because there was only one competitor."

They chose well: Carriage House has developed a devoted following as one of the best apple brandies being made in the southern craft world, and it's considered one of the best products being made in North Carolina.

When people ask what Nordan does, he tells them it's so weird, they'll never believe it: he flies hot-air balloons (true), builds furniture (true), and owns a distillery. The last one is always the one they never believe.

Carolina Distillery goes through 200,000 pounds of North Carolina-grown apples every year for its brandy, made by master distiller Tim Sisk, "probably the finest liquor maker in the state, legal and illegal," Nordan says.

It's aged in white oak barrels Nordan is having made in Kentucky from North Carolina lumber.

Building sales for the product, Nordan goes all over the state visiting customers, often hauling empty barrels that ABC stores use for displays. He loves to head to the Outer Banks, pulling his truck onto the ferry for the crossing. When fellow riders see the barrels, they always want to know if there's liquor in them. When he says no and their faces fall, he'll say, "But I've got some in the truck." He's made friends all over the state that way.

10   ASHEVILLE DISTILLING CO./TROY & SONS

12 Old Charlotte Hwy., Asheville; 828-575-2000, www.ashevilledistilling.com. Tours at 5 and 6 P.M. Friday and Saturday; free, no reservation required. Private tours by appointment.

Troy Ball has attracted attention in the distilling world for several reasons. An elegant, blonde businesswoman and long-distance equestrian, she was the first female distiller in North Carolina and is still one of the few in the country. She named her distillery, Troy & Sons, for her three sons; two, Marshall and Coulton, have special needs and are nonverbal and confined to wheelchairs, while the youngest, Luke, is a college student who works with her in the distillery.

And finally, she's making serious moonshine from heirloom Crooked Creek corn raised on a 200-year-old farm.

Ball's story is recounted in her memoir, *Pure Heart: A Spirited Tale of Grace,*

Moving to Asheville, N.C., for her sons' health was the first step in Troy Ball's journey to become a whiskey maker.

Wit, and Whiskey, released in spring 2017. She and her husband, Charlie, a real estate investor, were raising their sons in Texas, but the heat and pollen were hard on Marshall and Coulton's health issues. In 2004, they sold their farm and moved to the North Carolina mountains for a better climate.

In Texas, Ball had devoted all her time to the boys' care, even helping to cofound a center in Austin for children with autism and developmental difficulties. In Asheville, with the boys' health doing better, she had more time to herself. When the Balls moved in, neighbors brought local, homemade moonshine as gifts. Most of it was terrible, but when Ball finally tasted a batch that was excellent, she got interested in learning how to make it.

The real estate crash of 2008 wiped out her husband's business, so she ended up partnering with a local farm and starting a distillery herself to support her family. In only a few years, her corn whiskeys, including a clear whiskey and a lightly aged Blonde Whiskey, have attracted medals and a following at local bars.

You probably won't meet Troy herself when you tour the distillery. Tours are only offered at 5 and 6 P.M. on Fridays and Saturdays, when she's busy with her family. The tour guide I met knew his subject well and gave good explanations. The distillery is located in a park-like setting at the top of a steep, winding road, in the same building as the much larger Highland Brewing. Highland owner Oscar Wong, considered the grandfather of Asheville's craft brewing world, is a supporter and fan of Ball. A stop at Asheville Distilling is a good chance to combine tours of beer and whiskey for a "shot and a chaser" experience.

11 MUDDY RIVER DISTILLERY

1500 River Dr., Belmont; 704–860–8389, www.muddyriverdistillery.com. Short tours (30 minutes) and long tours (1 hour 15 minutes) are both $8; hours vary and are posted on the website, usually for Saturday afternoons and Tuesday evenings.

If you want to dip a paddle in the water, you actually can kayak up to Muddy River: it's in a brick complex that used to house textile mills, overlooking the Riverside Marina on the Catawba River about 30 miles west of Charlotte.

Owner Robbie Delaney and his wife, Caroline, followed their own strange path to running a rum distillery in the small town of Belmont. Delaney was in construction, and when the economic downturn slowed building, he had to travel farther afield to find jobs. In 2011, he was commuting between Texas and North Carolina when he saw an article in an in-flight magazine about the growth of craft brewing, with speculation that the next big thing would be distilling.

"I thought you had to be a billionaire, you had to be Bacardi, to make rum."

He started making it at home with a friend who often crashed on his couch. Since his friend was a fan of Captain Morgan spiced rum, they tried it.

"The worst rum ever," he says. "Our buddies were like, 'Don't quit your day job.'"

But he wanted to quit his day job. So he built a bigger still in his parents' backyard, figuring that bigger batches would help him make better rum. They threw him off the land, trying to discourage him. It was too late: he was already hooked.

Robbie Delaney of Muddy River Distillery in Belmont, N.C., got bit by the rum bug while reading an in-flight magazine.

"I became obsessed," he says. "You put stuff in one end and liquor spits out the other end."

He and Caroline decided to do it, finding a building on Craig's List, wrestling with state and federal regulations and finally opening as the fifth legal distillery in the state.

Delaney gives a great tour, showing off his stills, Democracy and Liberty (a third, Freedom, had been set up as a display when I was there, so you could look inside and see how the plates in a continuous still work). He even pours liquor on a steel work table and sets it on fire to make one point.

Inside the barreling room for your tasting, you'll see long stacks of charred oak barrels holding his Queen Charlotte's Reserve rum, which is getting good reviews. He's tapping every other barrel, letting some age as long as possible. One barrel was put down for the birth of a friend's daughter, with plans to open it on her twenty-first birthday.

They also make a spiced rum that's a lot better than his first attempt, flavored with sarsaparilla, vanilla, cinnamon, and gentian root. And give him props for making a coconut rum that doesn't taste like suntan oil: he uses fresh, young coconuts, so it tastes like real coconut. Mixed with pineapple juice, it makes a great beach quaff—or something to pack on your kayak.

12   DOC PORTER'S CRAFT SPIRITS

232 E. Peterson Dr., Charlotte; 704-266-1399, www.docporters.com. Longer owner-guided tours at 4:00 and 5:30 P.M. Saturdays are $15, and mini tours and tastings 6–8 P.M. Fridays and 2–7 P.M. Saturdays are $7.

There is an actual Doc Porter behind Doc Porter's, and there is a family behind this small distillery off South Boulevard in Charlotte.

Andrew and Liz Porter are the young couple who opened it, and they took the name from Andrew's grandfather, Dr. Richard Porter, who was a radiologist. He died just before their wedding, and when they started the distillery a year later in 2015, they decided to name it in his honor.

Liz is in marketing and Andrew is a chemical engineer. Both haven't quit their day jobs, running the distillery on the weekends with the help of a distiller who does the alcohol runs during the week.

It's a full grain-to-glass operation, with corn and wheat coming from a local farm and barley that's malted in Asheville, two hours away. (The orange and lemon peels used in their vapor-infused gin even come from Living Kitchen, a raw-food restaurant in Charlotte's nearby South End district.)

"Anything that comes out of Doc Porter's is really and truly from scratch," Liz Porter says.

They started with vodka and gin, then started releasing their bourbon, aged for 9 months in fifteen- and thirty-gallon barrels, a year later. They're also working on a rye and plan to add both absinthe and amaro eventually.

With his background in chemistry, Andrew Porter leads a solid tour, including passing around glass vials of heads, hearts, and tails, and even pulling a laboratory glass of sour, chunky fermenting corn from the fermentation tanks and passing it around to smell and taste.

One thing that makes the tour stand out is more attention to and explanation of vodka. While a lot of distilleries just make vodka as a starting point and don't give much attention to it in tours, Porter explains things like the proofing tank, where they lower the alcohol content with water and filter it through charcoal. Made with the traditional North Carolina soft red winter wheat, it has a distinctive butterscotch note.

"It's very difficult to be a small distillery because of the way we're taxed [in North Carolina]," he says. "But being small, we're closer to our product and our customers.

"The culture of Charlotte really is receptive to craft."

13 GREAT WAGON ROAD DISTILLING COMPANY

227 Southside Dr., Charlotte; 704-489-9330, www.gwrdistilling.com.
Tours on the hour 1–4 P.M. Saturdays; $5.

In southern history, the Great Wagon Road was the path that brought settlers, particularly Germans, from Pennsylvania down into western North Carolina. Their mark is still all over the North Carolina Piedmont, from the distinctive Lexington barbecue style, believed by some to have German origins, to family names like Oehler and Stamey. So it's a little ironic that the Great Wagon Road Distilling Company is distinctly Irish, not German.

Located in Charlotte's Lower South End district not far from uptown, it fits right in with the city's fast-growing brewery scene. It's actually in a small building on the same busy street with the large Old Mecklenburg Brewing beer garden down the block and another craft brewery, Sugar Creek, next door.

Great Wagon Road is a double experience: although North Carolina has rules against combining a bar and distillery, owner Ollie Mulligan, a native of County Kildare, Ireland, has gotten around it by having a bar, the Broken Spoke, under separate management next door. You go in there to arrange your tour and get a cocktail crafted with Great Wagon Road's Irish-style whiskeys, then go outside and enter through a separate door for the short tour of the distillery.

If you need to kill time waiting, there's a green-and-white Irish "Telefon" booth inside the bar. Enter a code that's on a nearby chalkboard and you can make a free call to Ireland.

The distillery tour is short and a bit perfunctory, with a brief explanation of the 3,000-liter column/reflux still and the fermentation tanks. The real draw is the tasting of 100 percent barley Irish-style whiskeys and an explanation of poitín, aka poteen, Ireland's version of moonshine. Mulligan's grandfather was a poitín maker with a group of friends who would split up the pieces of their still to keep them hidden. One night while riding his bicycle home with the condenser tucked under his coat, he was stopped and arrested. He claimed he was only making medicinal whiskey and got his fine reduced from 200 pounds to thirty-six pounds.

Great Wagon Road is producing several spirits, including a wheat vodka, Ban, a poteen, Drumlish, and a barrel-rested poitín, Quinn's. The highlight, though, is Rua, a barrel-aged Irish-style whiskey that was named the Best Beverage in the South by *Garden and Gun* in 2016.

Other Craft Distilleries in North Carolina

Blue Ridge Distilling Co. 228 Redbud Lane, Bostic; 828-245-2041, www.defiantwhisky.com. Makers of Defiant, an American single-malt whiskey. Tours 9 A.M.–3 P.M. Monday–Friday, weekends by appointment.

Broad Branch. 756 N. Trade St., Winston-Salem; 336-602-2824, www.broad branchdistillery.com. Tour and tasting at 6:30 P.M. on the first and third Friday of each month; appointment required; $5.

Covington Vodka Distillery. 310 Kingold Blvd., Snow Hill; no phone number available, www.covingtonvodka.com. Makers of sweet potato vodka. Free tours are available 9 A.M.–5 P.M. Monday–Friday.

Diablo Distilleries. 245 Jim Parker Rd., Jacksonville; 910-545-7010, www.diablodistilleries.com. Tours by appointment.

Fair Game Beverage Company. 193 Lorax Ln., Pittsboro; 919-245-5434, www.fairgamebeverage.com. Tasting room hours: 4–8 P.M. Friday, 1–7 P.M. Saturday, and 1–5 P.M. Sunday. Tours: 6 P.M. Friday, 1:30 and 3:00 P.M. Saturday, and 12:30 P.M. Sunday. Tours are $10 and tastings are $5; both include a souvenir glass.

Seventeen Twelve Southern Spirits. 300 Thornburg Dr. SE, Conover; 828-381-2949. Tours at noon, 1 P.M., and 2 P.M. Saturdays.

Greensboro Distilling Company/Fainting Goat Spirits. 115 W. Lewis St., Greensboro; 336-273-6221, www.faintinggoatspirits.com. Tours are offered from 3–7 P.M. Fridays, 1–6 P.M. Saturdays, and 1–5 P.M. Sundays for $10, including a souvenir glass.

H&H Distillery. 204 Charlotte Hwy., Asheville; 828-338-9779, www.hhdistillery .com. Makers of H&H Hazel 63 Rum. Tours 2–5 P.M. Thursday and Friday, noon–6 P.M. Saturday. Call or schedule online.

Howling Moon Distillery. 42 Old Elk Mountain Rd., Asheville; no phone number available, www.howlingmoonshine.com. Free tours by appointment. Email: cody@howlingmoonshine.com.

Mystic Farm & Distillery. 1212 North Mineral Springs Rd., Durham; no phone number available, www.whatismystic.com. A farm-to-flask distillery making Mystic Bourbon Liqueur and a single-barrel bourbon, Heart of Mystic. Tours at 2 and 4 P.M. most Saturdays.

Outer Banks Distilling. 510 Budleigh St., Manteo; 252-423-3011, outerbanks distilling.com. Makers of Kill Devil Rum. Tours at 1 and 3 P.M. Tuesday–Saturday; $10, reservation required.

Raleigh Rum Company. 1100 Corporation Pkwy., #132; no phone number available, www.raleighrumcompany.com. Free tour and tasting at 2 P.M. Saturdays. Email info@raleighrumcompany.com for details.

Seventy Eight °C Spirits. 2660 Discovery Dr., #136, Raleigh; 919-615-0839, www.78cspirits.com. Making several versions of limoncello. Tours by appointment.

Southern Artisan Spirits. Kings Mountain; 828-773-7536, www.southern artisanspirits.com. One of the first craft distilleries in the state with their Cardinal Gin and barrel-rested gin, SAS had not yet completed the store and tasting room when we checked. It's worth keeping an eye out for the future.

Southern Distilling Company. 211 Jennings Rd., Statesville; 704-978-7175, www.southerndistillingcompany.com. A large distillery that includes a demonstration farm, an orchard, and mini-donkeys. Tours and tasting by appointment on their website; $15.

Southern Grace Distilleries. 130 Dutch Rd., Mount Pleasant; 704-622-6413, www.southerngracedistilleries.com. A distillery located in an old prison that was built in 1929. Tours of both the whiskey operation and the prison are $14. Tours are 12:30 P.M., 2:00 P.M., and 3:30 P.M. Fridays and Saturdays.

Sutler's Spirit Co. 840 Mill Works St., #120, Winston-Salem; 336-565-6006, www.sutlersspiritco.com. Tours offered twice a month on Friday or Saturday from 6 to 9 P.M.; $10. Book reservations online.

Tryon Back Door Distillery. 11 Depot St., Tryon; 864-237-1667, www.tryon backdoordistillery.com. Tasting room open 11 A.M.–6 P.M. Tuesday–Saturday.

Walton's Distillery. 261 Ben Williams Rd., Jacksonville; 910-347-7770, www.waltonsdistillery.com. Free tours and tastings from 10 A.M.–5 P.M. Monday–Saturday.

Recipes

RED ROVER, RED ROVER

From Bob Peters, the beverage manager of the Punch Room at the Ritz-Carlton, Charlotte. While Peters likes to use Cardinal Gin, made by Southern Artisan Spirits in North Carolina, he says any American-style gin will work.

MAKES 1 DRINK

1 lime wedge
2 (1-inch) cubes seedless watermelon
2 ounces American-style gin
2 ounce salted basil simple syrup (recipe below)
Pickled watermelon rind (optional; garnish)

Place the lime wedge and watermelon in a shaker and muddle (crush). Add the gin and the salted basil simple syrup. Shake.

Fill an old-fashioned or rocks glass with ice. Double-strain the contents of the shaker over the ice and garnish with the pickled watermelon rind.

Salted Basil Simple Syrup
Place 2 large handfuls of roughly chopped sweet basil leaves and 4 cups of water in a saucepan. Bring to a boil, then reduce heat and simmer for 20 minutes. Add 1 cup of sugar and 1 tablespoon of salt, stirring until dissolved. Remove from heat and strain, discarding the basil leaves. Cool and store in the refrigerator for up to 2 weeks.

COPPERHEAD SHINERITA

From George Smith, Copper Barrel Distilling, North Wilkesboro, N.C. Moonshines macerated with fresh fruit were called bounces. They can look a little strange: the alcohol removes the color from the fruit. Most makers strongly suggest you don't eat the fruit itself. It can be unpleasantly strong.

MAKES 1 DRINK

2 or 3 slices of fresh jalapeño, seeds and ribs removed
1 1/2 ounces strawberry-flavored moonshine (preferably flavored with fresh fruit, not flavoring syrup)
Bottled lime margarita mix, such as Jose Cuervo
1 fresh strawberry

Place the jalapeño slices in a rocks glass and muddle (crush) gently. Top with ice, moonshine, and margarita mix. Garnish with the strawberry.

MOONLIGHTING IN BERMUDA

From George Smith, Copper Barrel Distilling, North Wilkesboro, N.C. If you tour enough distilleries, you'll inevitably end up with some unaged corn whiskey, whether it's labeled moonshine, white lightning, or corn whiskey.

MAKES 1 DRINK

1 1/2 ounces unaged corn whiskey or moonshine
1 1/2 ounces fresh apple cider
Ginger beer, such as Barritts
Cinnamon stick (optional, garnish)
Apple slice (optional, garnish)

Fill a rocks glass with ice. Add the corn whiskey and apple cider, then top off with ginger beer. Garnish with the cinnamon stick and apple slice.

THE SERENADE

From Kevin Gavagan, Charlotte, N.C. Kevin and I bonded over a mutual love of craft cocktails. Kevin and his wife, Heather, put on cocktail popups around Charlotte under the name Haunt Bar, while they wait to someday open their dream bar. At a chef event in Charlotte, Kevin introduced me to this cocktail, made with Copper Barrel moonshine. It was a big moment for me: the first time I'd had a cocktail made with moonshine that didn't overwhelm the flavor.

MAKES 1 DRINK

1 ¹⁄₂ ounce good-quality moonshine, such as Copper Barrel
1 ¹⁄₂ ounce Cantaloupe Shrub (recipe below)
Small handful cilantro
Ice

Muddle the cilantro in the bottom of a cocktail shaker. Pour in the shrub and moonshine. Add ice and double strain (see note) into an old-fashioned or rocks glass. Garnish with a cantaloupe chunk from the shrub and a cilantro sprig.

NOTE: Cocktails that are muddled with herbs need to be double-strained to get all the little bits of green matter out. To do it, hold a small fine-mesh kitchen strainer over the glass and the pour the contents from the shaker through both the strainer on the shaker and the kitchen strainer into the glass.

Cantaloupe Shrub

Cover about 4 cups of peeled, chopped cantaloupe with ¼ cup simple syrup and refrigerate overnight. Strain the syrup, setting aside the cantaloupe for garnishes, and add about 4 tablespoons white balsamic vinegar to the shrub.

A Different Beast: Barrel-Rested Gin

With so many craft distilleries establishing their market niche in creativity, you can find bottles that Old Mr. Boston never dreamed about: quinoa whiskeys, triple-smoked ryes, aquavit macerated with oyster shells, red and green absinthes. There are even once-vanished varieties being rediscovered, like the sweetened Old Tom gin.

That's part of the fun of distillery tours: encountering things that are completely new. In the excitement of the moment on a distillery visit, though, you can find yourself coming home with liquors and then facing the question, *What the heck do I do with this*?

When I first began tracking these new styles, I found myself particularly puzzled by barrel-rested gins. Yes, they have juniper and a range of botanicals like a gin. But they also have the brown color and oaky flavor notes of a whiskey or bourbon. So what direction do you go in trying to come up with a cocktail for them?

"There's a hundred different ways to skin a cat," says Bob Peters, the beverage manager at the Punch Room at the Ritz-Carlton in Charlotte. Peters has developed a passionate following among craft-cocktail fans in North Carolina for his evil-genius ways backed with a solid grounding in the classics.

With barrel-rested gins, he says, you can think about it like a gin or like a whiskey.

"The plug-and-play method," he calls it. "Or, and I didn't come up with this one but I've heard people use it, 'The Mr. Potato Head method.' If you change the eyes out, you've got something different, but you know it's going to work. Instead of a straight gin cocktail, take something like a Negroni. It adds a ton of depth to a Negroni. Or take a traditional whiskey drink, like a Manhattan. It's fantastic."

He thinks of barreled gins as a little more complex and seasonal.

"I like to think about barrel-aged gin as a winter gin. This has got a little body and a little more depth." That's why the aged gins can often be better matches for those heavier bourbon cocktails we go to in winter.

"That way, it doesn't remind your palate of that summer gin cocktail. So it allows you to go to a different place in your head."

South Carolina

LIQUOR TRAIL

3

26

385

85

77

20

Columbia

Greenville

GEORGIA

NORTH
CAROLINA

95

SOUTH
CAROLINA

26

4 6
5
● Charleston

COMPARED TO THE RAPID GROWTH in some of the states around it, South Carolina got a late start in the distilling world. While it once had just as much bootlegging activity as North Carolina, particularly in the legendary "Dark Corner" area in the northwest, the state had restrictive and often convoluted laws on the sale of alcohol for a very long time.

The state's "blue laws," which once prohibited any retail sales on Sunday for liquor or anything else, stayed in effect in many places into the 1990s. Those were finally lifted for the most part, allowing stores to open after church, but you can still only buy liquor on Sundays in ten counties that allow variances. Until 2006, bars and lounges could only sell drinks made from 1.75-ounce mini bottles. (Ironically, since the bottles hold a little more than the 1.5-ounce shots used in most cocktails, that actually meant drinks in South Carolina were stronger than in other places.)

Much of the credit for easing the laws and making a craft distilling industry possible goes to Firefly Distilling on Wadmalaw Island, where a muscadine winery owned by Jim Irvin was producing so much of the sweet wine that he needed another use for it. Irvin's friend Scott Newitt, a beer and wine wholesaler based in Charlotte, North Carolina, suggested using the muscadine grapes to make vodka. But first, they had to persuade the state to lower the cost of a distilling license from a prohibitive $50,000 to $1,200. The result was Firefly's sweet tea–flavored vodka. By 2009, it was a runaway hit all over the Carolinas, and a bunch of other Firefly products followed.

The change came at a time when the economy was still tough, though, particularly for start-up businesses. It really took until 2011 before many distilleries were able to secure business loans and begin building.

Today, while South Carolina liquor stores aren't owned by the state as they are in North Carolina, they still have to be physically divided, with beer and wine reached through one door and the liquor section reached through another. (Since the state has very low taxes on alcohol and gas, an awful lot

James Craig does the honors by snapping a picture of a bridal party kicking off a big weekend at the Striped Pig Distillery in Charleston.

of people from North Carolina still make the trek over the border to fill their tanks and their trunks on the same trip.)

While it still has fewer distilleries than most of the states along the Eastern Seaboard, South Carolina is catching up fast. In particular, the food-crazy city of Charleston is seeing a lot of growth. And because of the interest in bringing back heirloom varieties of grains and produce, helped along by research into antique and heirloom plant strains by University of South Carolina English professor David Shields, some of the most serious distilleries are creating spirits using long-neglected crops, from Italian ryes that date to the colonial era to the legendary Bradford watermelon, a fruit prized for its flavor that fell out of favor because its rind was too delicate to handle shipping.

For this trail, we'll start in the northwest, outside Greenville, and work our way east, ending in Charleston.

1 SIX & TWENTY DISTILLERY
3109 Highway 153, Piedmont; 864–263–8312, www.sixandtwentydistillery.com.
Tours at 1 and 4 P.M. Saturdays; gift shop and tasting
room open noon–6:20 P.M. Monday–Saturday.

While the location of the Six & Twenty Distillery is officially in the town of
Piedmont, locals call the area Powdersville. Just to be a little more confusing:
it's actually on the outskirts of Greenville, a city that's developing a reputation for its lively food-and-drinks scene.

The distillery isn't fancy, in a cinderblock industrial building with a small
tasting room and a large warehouse area in the back that holds both the distillery and the barrels.

The co-owners, David Raad and Robert "Farmer" Redmond, played rugby
together at Clemson University before becoming businessmen. Redmond is
sort of distilling royalty: he's the great-great-nephew of Lewis Redmond, a
notorious figure in late nineteenth-century Appalachia.

In the years after the Civil War, Redmond was better known in American
folklore than Jesse James. A moonshiner and whiskey runner in the mountains of both North and South Carolina, Redmond became an outlaw after
shooting a federal agent who tried to arrest him for bootlegging. But he also
was known for sharing his money with poverty-stricken people in the hard-
scrabble region, sort of the Carolinas' version of Robin Hood. He appar-
ently was more valuable to locals than the officer he shot: when he was captured, he was only given a ten-year sentence and was pardoned by President
Chester Arthur after serving only three years. After he was released, he went
legal with whiskey in the final years of his life. Some sources claim he was the
last legal distiller before he died in 1906.

The Six & Twenty staff leads a good tour of the distillery, in a warehouse
behind the gift shop, passing around jars of heads, hearts, and tails so you can
smell them and learn the difference. Good to know: the tails smell like sweet
wet cardboard and taste like overcooked green beans.

The tasting is generous, covering their wheated whiskey Old Money;
5-Grain Bourbon from soft red winter wheat, corn, barley, rye, and rice; and
Carolina Rouge, a wheated whiskey aged in French Cote Rotie wine barrels.
The most popular product is Carolina Cream, sort of a bourbon version of
Irish cream. They also make an unusual brandy from Bradford watermelons.
When I was there, the label decoration was a tribute to the people who died

in the shooting at Emanuel African Methodist Episcopal Church in Charleston, and some of the profits were being used to help the families.

Where did the name Six & Twenty come from? According to local legend, a Choctaw princess, Issaqueena, had been captured by the Cherokee and overheard plans to attack a British fort where her lover, a soldier, was serving. She escaped and rode ninety-six miles to warn him, naming creeks and landmarks along the way with the distance so she could find her way back. That's also the source for a couple of town names in the region, Six Mile and Ninety Six. The distillery is near a creek Issaqueena called Six and Twenty, because it was twenty-six miles from the Cherokee village, which was located under what is now Lake Keowee.

2 DARK CORNER DISTILLERY

14 S. Main St., Greenville; 864-631-1144, www.darkcornerdistillery.com.
Self-guided tour and tastings $7, including souvenir glass, 11 A.M.–7 P.M.
Monday–Saturday (hours can change by season, so call ahead to check).

North Main Street in Greenville is popular for strolling and shopping, a busy strip of boutiques and jewelry shops. Dark Corner no longer distills in its small building—the stills and bottling have been moved to a larger location outside town.

The storefront downtown is worth a stop, though. It's been set up as a small museum, with displays of moonshining equipment and signs that do a good job of covering the basics.

At the tasting bar, you can try their moonshines and flavored moonshines, along with Whiskey Girl, a flavored whiskey; Hot Mama, with cinnamon and chipotle pepper; and a premium aged whiskey, Lewis Redmond Bourbon. See? Redmond is still a very popular guy in this part of the world.

3 COPPER HORSE DISTILLING

929 Huger St., Columbia; 803-779-2993, www.copperhorsedistilling.com.
Tours usually 6 P.M. Fridays, and 1, 4, and 6 P.M. Saturdays; free, including tasting.

When you write a book about fledgling businesses, their struggle is part of the story, too. Here's hoping that Copper Horse manages to hang on until owner Richard Baker gets a chance to tap all those barrels he has in his warehouse on the edge of downtown Columbia. From the steep front steps, you can see the dome on the Capitol and the tall tower of Adluh Milling, which supplies the South Carolina–grown grain he's using for his whiskeys.

Baker's operation is worth a stop for his devotion to the craft of distilling. But he's also worth talking to for his pragmatism about the difficulties facing craft distilleries.

"I don't consider myself a master distiller," he insists. "That's guys like Jimmy Russell [of Wild Turkey] and Fred Noe [of Jim Beam]. Maybe when I have twenty years under my belt."

Baker has degrees in biology and chemistry, but he spent twenty years working for his family's computer business. He had worked at a couple of distilleries for fun. Whiskey, he says, was both a fascination and a family history: "During Prohibition, what family didn't make it back in the day?"

In 2008, when his family sold the computer business, he had his chance. He was considering moving to the Bahamas, but a hurricane went through about that time, and he decided to stay put somewhere that wasn't likely to blow away. So he started his distillery, planning to go slow and keep his whiskey in barrels until he thought it was ready to release. But then the liquor business in South Carolina began to take off and he now finds himself facing a lot more competition than he had expected.

"Right now, bourbon is white hot," he says. For a one-man distillery, he says, it was tough enough just making enough product to sell. On top of that, though, he also has to fight for distribution and shelf space in liquor stores.

He worries that by the time his bourbon is ready to release—the oldest barrels were three years old when I met him—there will be too many small bourbons to find a market. To hedge his bets, he's added rum, gin, a whiskey-flavored cream, and novelty things like Hot, a dangerously hot vodka flavored with Carolina Reaper and ghost peppers.

He loves making whiskey, and his love for it and his knowledge of science combine to make an excellent tour.

Still, he admits that if he had the chance again, he wouldn't risk it.

"There's no glamour in this if you're not prepared to work constantly," he says. "We're all competing, we all want the chance.

"What other business is this difficult?"

4 STRIPED PIG DISTILLERY

2225-A Old School Dr., Charleston; 843-276-3201, www.stripedpigdistillery.com.
Tours 3–6 P.M. Thursdays and Fridays, noon to 4 P.M. Saturdays;
$5 including tasting. Private tours $15 by appointment.

Some guys have all the fun. At Striped Pig, in a warehouse in an industrial park north of downtown Charleston, Todd Weiss and James Craig can barely hold in their delight at their new business.

When I pulled in for a tour late on a Friday afternoon, a boisterous group of young women was piling in right behind me: it was a bridal party kicking off a celebration weekend, complete with plastic tiaras and a sparkly sash on the bride-to-be.

The only thing that could match their energy was Craig and Weiss themselves. Pretty soon, the bride was stretched out on the tasting bar and Craig had grabbed her camera to shoot pictures.

Weiss was in sports medicine and Craig was selling medical equipment, but they had never come across each other before Weiss got interested in distilling. Weiss's father had owned a home-brewing supply shop and his brother was a brewer.

"I was the black sheep," he says. "I liked liquor."

Weiss wanted to try making liquor, so he and his friend Johnny Pieper put an ad on Craig's List looking for an old pot still. Craig saw the ad and got so curious he called just to ask why they wanted one. He came out to see what they were building and ended up joining the business.

They opened in 2013, starting with white rum and then adding bourbon, gin, corn-based vodka, and corn whiskey. Craig handles the sales and marketing, and they all do the distilling. They're enthusiastic entertainers, renting out space in their big warehouse for parties and art shows. (If you go in the summer, be prepared: it's air-conditioned inside the tasting room, but the warehouse can get very hot. It is Charleston, after all.)

The name Striped Pig came from an old Charleston story: before Prohibition, rum sellers in Charleston couldn't sell less than fifteen gallons of liquor, in casks called hogsheads. An industrious vendor put up a tent with a flag that showed a picture of a hog's head and an offer to see his "striped" pig (disguised with paint). If you paid 6 cents and stepped inside to view the pig, you got a "free" sample of rum.

Craig has all the personality of a natural-born salesman, so he usually handles the tours. (He's also known for goofy stunts like taking the distillery's pet pig, Jackson, for walks on a leash in downtown Charleston.) If you ask Craig anything, he'll answer, "Let me tell you a quick story," and Weiss will roll his eyes and warn you: "Jim never tells a quick story."

5   CHARLESTON DISTILLING CO.

501 King St., Charleston; 843-718-1446, www.charlestondistilling.com.
Tours 11 A.M.–7 P.M. Monday–Saturday (closed Wednesdays); $5.

In Charleston's historic downtown, if you get tired of the high-priced boutiques and the even higher-priced sea-grass baskets, you can stop by Charleston Distilling, in the middle of everything on King Street.

Their ryes, bourbon, gins, vodka, and an unusual ginger cinnamon liqueur are all made here, in a large facility with a comfortable tasting room that's set up to handle large groups. (Check out the restrooms, built into replicas of giant barrels.)

All the grains, including Abruzzi rye and red winter wheat, are grown in nearby Summerville, and the tour is a good, if basic, look at the distilling process.

One thing you won't see there: barreling. The barrels are being aged in an unheated warehouse out on Johns Island, the farming area about ten miles southwest of the city. Charleston real estate is too expensive to allow that much space, and the warehouse isn't a good neighbor: they're doing "audio aging," playing loud music around the clock to cause vibrations in the barrels that are believed to hurry the alcohol's contact with wood. They've been playing the same Mozart CD on full blast for two years.

Employees at the distillery admit they try to avoid getting assigned to warehouse duty; it's very hot and very loud.

6   HIGH WIRE DISTILLING CO.

652 King St., Charleston; 843-755-4664, highwiredistilling.squarespace.com.
Tours hourly from 11 A.M.–6 P.M. Tuesday–Saturday; $8 with complimentary tasting (a tasting flight of up to three ounces).

From its eye-catching labels, with colorful images of magic acts—founder Scott Blackwell was a magician as a kid—to the unusual list of liquors based on heirloom produce, High Wire is the cool kid on the block in Charleston, a city that's almost obsessed with food and drink.

Hat Trick, a gin flavored with South Carolina–grown botanicals, is one of the most popular offerings at High Wire Distilling in Charleston.

Nick Dowling, the production distiller and one of only a handful of employees, showed me around for my tour. Located in a warehouse in the section of the city Charlestonians call "Upper Upper King," it's next door to a bakery and coffeehouse, and its tasting room has become a popular hangout and event space.

Everything is made in Nadine, "our fifth employee," a hybrid pot/column still that allows a lot of flexibility, moving the plates inside around to get the effect they want. (The still was named Jennifer when they bought it from Troy & Sons in Asheville, but they wanted a name that was more southern and "Charleston.")

Blackwell, who owns High Wire with his wife, Ann Marshall, trained at the Culinary Institute of America as a baker and founded an organic bakery, the Immaculate Baking Co., which he later sold to General Mills. He describes his distilling style as "culinary," focusing on unique local ingredients.

Many of the ingredients they work with are in collaboration with David Shields, the University of South Carolina food historian who has made a career of rescuing old plant varieties. Their New Southern Revival whiskeys include a sorghum whiskey, a wheated rice bourbon made with Carolina Gold rice and a rye made with Abruzzi rye that was brought to the area by Italian immigrants.

Their Southern Amaro, an Italian-style bitter and sweet liquor, uses black tea from a local tea plantation, along with mint, Dancey oranges, and Yaupon holly, a popular southern shrub that's the only caffeinated plant native to North America.

"We're somewhat obsessed with it," Dowling says of Yaupon. Its leaves were used by Native Americans to make a tea that's similar to the yerba maté used in South American cultures.

Dowling gets very excited about all the experimenting, particularly with the things Shields brings to their attention, such as Bradford watermelons for a small batch of watermelon brandy and a bourbon they're experimenting with made from Jimmy Red corn.

"It is so different," Dowling told me. "Buttery, rich, oily." When they mashed it, it developed a one-and-a-half inch cap of red corn oil on top of the mash.

For a lot of fans, the favorite is Hat Trick gin, a New West style flavored with juniper, lavender, orange and lemon peels, lemon grass, licorice, angelica root, coriander and cardamom, then barrel-rested for six months in virgin oak barrels.

Being a small distillery in a city filled with hot restaurants and serious cocktail programs means they can get creative making things that are becoming signature liquors in Charleston.

"We're all little renaissance people here," Dowling says.

TRAVELER'S NOTE: Navigating Charleston's tangle of streets and finding parking can be notoriously difficult. There's a free trolley for locations along King Street (although not "Upper Upper King," where High Wire is located). Ride-sharing apps like Lyft and Uber make it much easier to get around and usually aren't all that expensive since most locations in the historic area are close together.

Carolina Moon Distillery. 116 Court House Sq., Edgefield; 803-275-7952, carolinamoondistillery.com. Tasting room hours: 11 A.M.–5 P.M. Tuesday–Saturday; free tours.

Copperhead Mountain Distillery. 14 S. Main St., Traveler's Rest; 864-610-2228, copperheadmtn.com. Small, family-owned moonshine distillery with a gift shop. Open noon–6 P.M. Tuesday–Wednesday, 11 A.M.–7 P.M. Thursday–Saturday.

Crouch Distilling. 947 S. Stadium Rd., Columbia; 803-764-6839, www.crouch distilling.com. Tasting room open 4:30–7:00 P.M. Thursday–Friday, 1–7 P.M. Saturday.

Daufuskie Island Rum Company. 270 Haig Point Rd., Daufuskie Island; 843-342-4786, www.distillerysolutions.com/daufuskie/tours.php. 10 A.M.–4 P.M. Wednesday–Saturday. Tour and tasting is $7.

Firefly Spirits. 6775 Bears Bluff Rd., Wadmalaw Island; 843-557-1405, fireflyspirits.com. The first distillery to open in South Carolina in modern times with sweet tea vodka; it now also makes whiskeys, bourbons, rums, and liqueurs. Open for tastings 11 A.M.–5 P.M. Tuesday–Saturday; $6.

Gorget Distilling. 1974 Whiting Way, Lugoff; 803-626-0077, www.gorget distilling.com. 1–6 P.M. Thursday–Friday, 10 A.M.–4 P.M. Saturday; call for tours.

Hilton Head Distillery. 14 Cardinal Rd., Hilton Head Island; 843-686-4443, www.hiltonheaddistillery.com. Noon–6 P.M. Monday–Saturday. Tours for $20, tasting only for $10.

JAKAL Distillery. 106 Fabrister Ln., Lexington; 803-520-8323, www.jakal distillery.com. 10:30 A.M.–6:00 P.M. Saturdays; call for tours.

Lucky Duck Distillery. 17B Yemassee Hwy., Yemassee; 843-812-8337 or 843-589-5440, www.luckyduckdistillery.com. 10 A.M.–5:00 P.M. Tuesday–Saturday. Tours by appointment.

Motte & Sons Bootlegging Co. 220 E. Daniel Morgan Ave., Spartanburg; 864-308-1844, www.motteandsons.com. 11 A.M.–7 P.M. Monday–Saturday; free.

Straw Hat Distillery. 1301 N. Douglas St., Florence; 843-453-1349, www.strawhatdistillery.com. 9 A.M.–5 P.M. Saturdays; schedule tour by appointment online.

Sugar Tit Moonshine Distillery. 330 Main St., Reidville; 864-249-6483, www.sugartitmoonshine.com. 10 A.M.–7 P.M. Monday–Saturday; free tour and tasting.

Recipes

CHARLESTON STORM WARNING

From Striped Pig Distillery, Charleston. They like to use Cannonborough Soda ginger beer, a local brand, although Barritt's would work well, too.

MAKES 1 DRINK

2 ounces spiced rum
4 ounces ginger beer
Lime wedge

Fill a tall glass with ice, add the spiced rum and ginger beer and stir gently before serving. Top with a wedge of lime.

CRANBERRY GIN SMASH

From Scott Blackwell at High Wire Distilling, Charleston.

MAKES 1 DRINK

1 ounce Cranberry Spice Simple Syrup (recipe below)
1 ounce fresh lemon juice
2 ounces gin, preferably Hat Trick Botanical Gin
Cranberries and a rosemary sprig (garnish)

Fill a mixing glass halfway with ice. Add the simple syrup, lemon juice, and gin. Cover with the bottom of a cocktail shaker and shake well for 10 to 15 seconds.

Using a fine mesh strainer, strain into a rock glass with a couple of cubes of ice or serve neat in a coupe-style glass.

Garnish with a rosemary sprig, skewered cranberries, and orange peel.

Cranberry Spice Simple Syrup

1 cup water

2-inch piece of orange peel (white pith scraped away)

1 cinnamon stick or 1 teaspoon ground cinnamon

10 cranberries

1 cup sugar

3 sprigs fresh rosemary

Place the water, orange peel, cinnamon, and cranberries in a small saucepan and bring to a simmer over medium-high heat. Reduce heat and simmer for 30 minutes. Remove from heat and cool briefly. Strain through a strainer, then add the sugar. Stir until dissolved, about 1 minute. Add rosemary and cool completely. Remove rosemary before using. Refrigerate for 2 to 3 weeks.

The Dark History
of Dark Corner

While North Carolina is known for its rugged, wildly beautiful mountains and legends of the rough lives early settlers lived there, South Carolina has its own mountain region, not as steep as the Smokies but rugged, beautiful, and once just as remote.

The area around Landrum and Glassy Mountain has been nicknamed the "Dark Corner" since the nineteenth century. While it had plenty of moonshining, bootlegging, and an outlaw reputation, that wasn't the source of the name.

In 1832, the state of South Carolina went to battle against the U.S. government, claiming that two tariffs, passed in 1828 and 1832, were unconstitutional. The state passed the Nullification Act, declaring the tariffs null and void in South Carolina. A small area in the northwest was the only district that voted against it, causing Vice President John C. Calhoun, an S.C. native, to declare that the area was "a dark corner where the light of Nullification would never shine."

After the Civil War, the area gained a reputation for resisting authority and pursuing occupations like moonshine in a place that was remote and difficult to access. Eventually, the whole northwest corner of South Carolina ended up being called the Dark Corner.

Where is it exactly? When outsiders would ask, local people used to say "just a little piece farther down the road."

While it doesn't have historical markers and most of the references to it now are in relation to distilleries, a good area to explore for the history of the region includes the Cherokee Foothills Scenic Highway as it crosses the North Carolina–South Carolina border.

Kentucky and Tennessee

LIQUOR TRAIL

4

Kentucky and Tennessee

1 Barrel House Distilling Co., Lexington, Ky.

2 Old Pogue Distillery, Maysville, Ky.

3 New Riff Distilling, Newport, Ky.

4 Kentucky Peerless Distilling Co., Louisville, Ky.

5 Copper & Kings, Louisville, Ky.

6 Limestone Branch Distillery, Lebanon, Ky.

7 Corsair Distillery, Bowling Green, Ky.

8 Corsair Distillery, Nashville, Tenn.

9 Nelson's Green Brier Distillery, Nashville, Tenn.

10 Tenn South Distillery, Lynnville, Tenn.

11 Prichard's Distillery, Kelso, Tenn.

12 Knox Whiskey Works, Knoxville, Tenn.

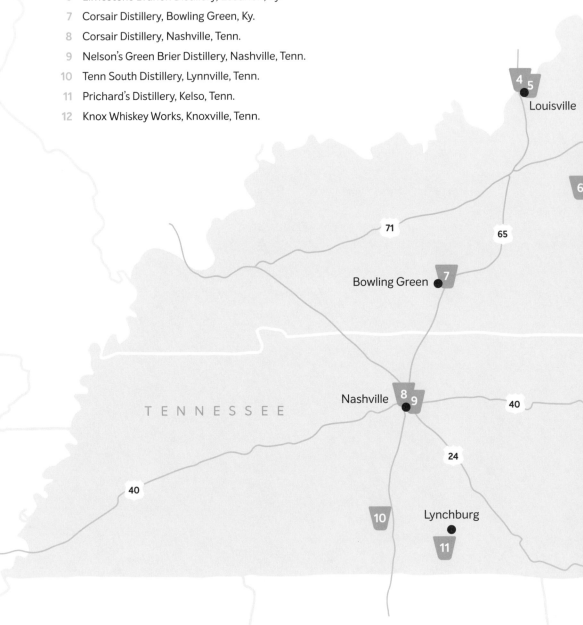

OHIO

3 Cincinnati

2

WEST
VIRGINIA

71 Lexington

1

KENTUCKY

75

VIRGINIA

81

Knoxville 12

NORTH CAROLINA

CRAFT DISTILLERS IN these states live in the shadows of giants. So many of the names that dominate American whiskey—Beam, Daniel, Sazerac, Van Winkle—grew out of these hills. Or more specifically, they grew out of the region's history, helped by excellent conditions for growing corn and pure water filtered by the limestone rock that rises out of the landscape, sometimes lining the roadways like fortress walls.

Sure, you can have a great time touring the major distilleries. Kentucky in particular has made it easy, establishing a Bourbon Trail across the state and even an Urban Bourbon Trail through Louisville. Whether you stick to the sanctioned trail or not (if you do follow it, I'd suggest getting off to include some fine distilleries that aren't in the official list), you'll see and experience a lot in efficient tours on gorgeous property. Woodford Reserve's manicured lawns, Maker's Mark's rustic homestead look, and Four Roses' Spanish mission style are great places to spend whole afternoons.

However, if you skip the official routes and visit these two states through the lens of small, startup distilleries, you'll get a very different and enlightening experience. You'll learn about the tough economic realities of small business, meet people who have to balance passion with pragmatism, and discover creations that are more experimental than most other areas. After all, if your neighbor is Jack Daniel's, why not do like Corsair in Bowling Green, Kentucky, and Nashville and start playing with ingredients hardly anyone else is using, like quinoa and spelt?

Because there are so many big players that can influence state policies, the two states also are a bit more liberal in what they allow visitors to do at distilleries. You aren't limited in what you can buy, and the two states also compete with each other. (If you're into collecting great liquors, you'll find some very well-stocked party stores in Kentucky where you can fill your trunk with things you don't find in other places.)

Tennessee loosened up its distillery rules in 2009, allowing the fast growth

George Pogue (center) leads a tasting at a bar set up in the nineteenth-century house on the Ohio River where generations of his family once lived in the town that first put bourbon on the map.

of microdistilleries, and Kentucky added cocktail sales at distilleries because Tennessee did it first. A little competition can be a healthy thing.

While Kentucky's historic ties to distilling have been helpful to start-ups, Tennessee's rules since Prohibition had been much more strict, with the two biggest distilleries, Jack Daniel's and George Dickel, grandfathered in. In 2009, the state made a big change, passing a law that allows distilling in any county that already has liquor stores and liquor-by-the-drink. That's opened up big areas of the state, and every major city now has distilleries.

For this tour, we'll start on the eastern edge of Kentucky's whiskey-producing region, in Lexington, and work our way north, then west to Louisville, south to Nashville, and back east, ending in Knoxville. We'll include a stop near Lynchburg, Tennessee, just in case you can't resist a stop at that big Jack Daniel's factory on the hill.

1200 Manchester St., Lexington, Ky.; 859-259-0159, www. barrelhousedistillery.com.

Hours: noon–5 P.M. Wednesday–Sunday. Tours start hourly.

They take distilling seriously in Lexington: there's now a Distilling District, labeled with street signs, to mark the old industrial area on the edge of the city that was once known for manufacturing and vice. Today, the brick warehouses there are bubbling up again, with a brewery, a restaurant, small bars and entertainment venues and even a running trail. There are several distilleries in Lexington if you want to make a day of it, including Blue Grass and Town Branch closer to the downtown area. Since we're covering a lot of ground, though, we're just going to focus on one in the Distilling District, Barrel House. Started in 2008 and opened to the public in 2010, it was one of the first craft distillers in Lexington.

When you turn into the district, don't stop at the first Barrel House you see: that's an entertainment space that's not associated with the distillery. The place you want is toward the end of the street on the left. It looks like the warehouse it used to be, but when I was there, they were busy building a new cocktail lounge on one side, because Kentucky had just passed the law allowing cocktail sales at distilleries.

Barrel House has bona fides: it's on the site of the old James Pepper Distillery, which opened in 1878 and closed in 1958.

Robert Downing, the lead distiller, says the current business started with a poker game. Co-owners Jeff Wiseman and Pete Wright were childhood friends who used to do an annual "guys' trip" with their buddies. Wiseman is a neurologist and Wright was in construction. All the guys in the group were bourbon aficionados, and one night, they started talking about whether it would be possible to make their own whiskey, just for fun.

"They thought it would be a unique challenge," Downing says. "They wanted to show the average man could do something like that."

Today, Wiseman and Wright work with two distillers, Downing and his assistant, Chad Burns, to make 9,000 gallons of spirits a year, using locally grown, non-GMO corn and spring water they haul to the distillery in a small tank.

Their moonshine, Devil John, is named for Pete Wright's great-great-uncle, a Confederate soldier with a storied past: he got captured and was forced to switch sides to the Union to save himself from execution.

Working with cobbled-together equipment, including a repurposed milk chiller from an old dairy as a mash cooker—"When you're working on a tighter budget, you tap into your ingenuity," Downing admits—they're making Devil John moonshine and barreled moonshine, RockCastle bourbon, and Oak Rum. Rum is still less common among Kentucky distilleries— sugarcane mostly grows farther south. There are advantages to being in Kentucky, though: they're aging in used Buffalo Trace and Four Roses barrels.

One fun touch: there's a big world map on the wall that visitors have peppered with pushpins to show their homes. While millennials come on weekends while they're hanging out in the Distilling District, baby boomers are usually on longer excursions, from Indiana and Florida. Downing looks forward to both groups.

"It's like you take a little ownership when you've visited a place."

2   OLD POGUE DISTILLERY

716 W. 2nd St., Maysville, Ky.; 317-697-5039, www.oldpogue.com.
Tours by appointment at 11 A.M. and 2 P.M. Thursday–Saturday, 2 P.M. Sunday.

Of all the towns you can visit in search of distilleries, Maysville is one of the most interesting. Although the town doesn't play it up—I couldn't find a single historical marker to note it—it's considered the birthplace of bourbon because of its location on the Ohio River.

In the eighteenth century, before railroads and highways took over, shipping really was mostly by water. The most efficient way to move goods from the mountain South to the port at New Orleans was by river. Fed by the Allegheny River in Pennsylvania, the Ohio borders six states and stretches along the northern border of Kentucky, passing Cincinnati and Louisville before finally joining the Mississippi at Cairo, Illinois.

Maysville, in the northeastern corner of Kentucky, was once an important port, where the banks were low enough to build wharfs. Several large distilleries sprang up to take corn from nearby farms and turn it into corn whiskey that was stored in charred oak barrels, stamped with the county name, Bourbon, before they were loaded onto boats for the long water trip. One of the first recorded uses of the word "bourbon" to refer to the area's barrel-aged corn whiskey was in 1821, in a newspaper advertisement printed in the nearby town of Paris.

Today, Maysville looks like something from Mark Twain, with clapboard

Distiller George Pogue was a geologist who can tell you all about the importance of rock to the story of Kentucky bourbon.

houses clinging to steep banks along the river and a massive wall, painted with romanticized images of nineteenth-century life, that protects the town from floods and the noise of a busy railroad line on the other side.

Old Pogue is the only distillery in town now, and it's a tiny place on the grounds of the Pogue family property. Don't let its size fool you: what Pogue is making, led by distiller George Pogue, is a nine-year-old bourbon that can hold its own with anything in the state.

You'll find Old Pogue a few blocks from downtown, down a very steep driveway that leads to a gracious riverfront house built in 1845. The tour starts in the tiny distillery in a small building next to the house, where George Pogue is struggling to keep up with his own success.

In addition to Five Fathers, a single-malt rye named for the five previous distillers in the family line, all named Henry Edgar Pogue (I through IV), Pogue produces fifty barrels of whiskey a year that he ages in barrels kept in a private location in Bardstown. When Pogue restarted the family business a few years ago, he aimed for wide distribution in eighteen states with bottles selling for $35. My husband and I tasted it in a whiskey flight at a Louisville restaurant in 2012, and I'm kicking myself now for not buying it.

When the bourbon craze took off, Pogue was overwhelmed by the demand. When you age for nine years, you can't just ramp up production. He ended up raising the price to $110 a bottle, and you have to order in advance and pick up at the distillery in October and November. The only other way to taste Old Pogue is to make the trip to Maysville for a tasting and tour.

"I might stop sleeping," Pogue says. "Volume is the typical restriction on us all."

Pogue, thirty-one when I met him, was originally a geologist, giving him a good perspective on the role of Kentucky's limestone-filtered water, drawn from 400 feet below the distillery.

"Pretty much all the water we harvest comes in contact with the rock," he says.

Before Prohibition, the Pogue Distillery made fifty barrels of whiskey a day. While it got one of the coveted licenses to sell its stock as "medicine," it couldn't make more, wiping out its supply. Once Prohibition ended, it was too expensive to restart making something that needed to be aged for years before it could be sold, especially when the Depression started. The distillery closed in 1933, and the abandoned building burned down in 1973. The family lived in the homeplace until the 1950s, when the house was sold. It eventually became rental property and then sat empty for ten years.

But the last Henry Edgar Pogue turned out to be a packrat, who saved all the records from the distillery and his last bottles of whiskey. When George and his cousins used to get together for family gatherings, there were boxes of bottles that had been sitting there, unopened, for sixty years. They started opening the bottles just to see how they tasted. Some of it wasn't very good, Pogue says—but some of it was great.

They all started talking about trying to re-create their family's bourbon, just for fun. They worked with a lab to settle on a yeast that would match the best of the old supply. They actually had some of the original yeast, although Pogue says they couldn't risk using it: yeast mutates quickly and there's no telling what would happen if they tried it.

They started in Bardstown in 1995, just as a hobby. By 2004, with interest rising again in bourbon, they came back to Maysville, repurchased the old home place, and started over. While no one lives in the historic family house, the first floor has been turned into a tasting room, with gorgeous views of the Ohio River from the front porch.

With his science background, George Pogue does a fine job explaining both the process of distilling and the history of bourbon in this corner of the state. Despite the size of the business, he says the relationship between craft distillers and large, premium distillers in Kentucky is congenial, with the big companies doing what they can to support the startups.

"We only do about fifty barrels a year," he says. "Those guys are doing a thousand barrels a day. But we found our niche."

TRAVELER'S NOTE: Maysville is a small town, but there's a comfortable hotel, the French Quarter Inn, next to the railroad line along the river (ask for a room on the town side to avoid the noise). And there's another tidbit that makes the town fun: it was the birthplace of singer Rosemary Clooney and her brother, TV commentator Nick Clooney. Nick's son George wasn't born in Maysville, but he spent a lot of time there as a kid. When he married his wife, Amal, in 2015, he brought her to see his family's hometown. They ate dinner at Caproni's, the Italian restaurant near the French Quarter, if you feel like rubbing elbows with celebrity along with your red sauce. If you're more interested in bourbon experiences, stop by the bar at Chandler's, about two blocks from the French Quarter, and check the shelves behind the bar: you might be rewarded with a chance at a shot of Pappy Van Winkle.

3   NEW RIFF DISTILLING

24 Distillery Way, Newport, Ky.; 859-261-7433, newriffdistilling.com.
Hours: Noon–7 P.M. Thursday–Saturday, noon–4 P.M. Sundays.
Tours by appointment through website, Tuesday–Sunday, hours vary.

After a bucolic drive through the farmland of northern Kentucky, it's a surprise to approach Newport and suddenly see high bluffs topped with big buildings and towering skyscrapers rising up around you: that's Cincinnati, right across the Ohio River. Newport was once the place where people from Cincinnati came for cheap booze and other vices. Today, it's a thriving bedroom community. But it's still the place Cincinnatians come to do their liquor shopping.

Just before you cross the bridge that carries you into Cincinnati, you'll see New Riff on the right. A sparkly new building of glass and wood, it looks a

bit like a suburban health club. It's also easy to spot because it shares a parking lot with a huge discount party store, The Party Source. (The sign out
front proudly declares that it's the largest party store in the state. And Kentucky is a state that likes to party, so that's saying something.) There's a great
selection of liquors in there, particularly ryes and bourbons, if you want to
get in a little shopping. The location isn't a coincidence: owner Ken Lewis,
a former high-school English teacher, started out investing in liquor stores
in Louisville and then opened The Party Source before deciding to open his
own distillery. Under Kentucky law, you can't own both a distillery and a
retail store, so he sold the store to his employees, although he still owns the
brewery located inside.

Across the parking lot, he built a slick distillery with a glass tower on one
side to display the tall, copper column still. Lewis is aiming for growth: the
building includes a huge event space upstairs, including a space set up for
cooking and tasting classes. New Riff has its sights on becoming a tourism
destination, with room to accommodate large groups and tour buses.

The distillery opened in May 2014, and their O.K.I. bourbon is currently
coming from the big Indiana facility MGP (O.K.I. stands for Ohio/Kentucky/
Indiana). When they release their own bourbon, planned in September 2018,
they'll switch the name to New Riff bourbon. They're also making two unaged corn whiskeys and two gins, the floral/botanical Kentucky Wild and an
aged version made in used bourbon barrels.

4 KENTUCKY PEERLESS DISTILLING CO.

120 N. Tenth St., Louisville; 502-566-4999, kentuckypeerless.com.
Hours (April–October): 10 A.M.–5 P.M. Wednesday and Saturday,
10 A.M.–7 P.M. Thursday and Friday. Tours on the half hour, $20.

A quick drive from Newport brings you into busy Louisville, with its snake
of interstate ramps near the river. Tucked underneath, in an old industrial
area, Peerless has taken over a 120-year-old brick building that used to be a
tobacco warehouse.

For the Taylor family, it's not only a return to their family history, it's a
move into the big city.

"We're bringing back an old brand," says owner Corky Taylor. Taylor's
great-grandfather, Henry Kraver, owned one of the largest distilleries in the
state, in Henderson, Kentucky, two hours to the west. Before it closed in 1917,

it made 200 barrels of whiskey a day, some of served in Kraver's hotel, the Palmer House in Chicago.

Corky Taylor spent his life in financial services before he retired to Florida.

"Walked on the beach for a year and half," he says. "Most depressed I ever was in my life. I said, 'I have to go back to work.'" So he paired with his son Carson, a builder, to outfit the old building in Louisville with a throwback feel: the gift shop includes an old truck and lots of farm implements.

"We wanted to make it feel like you're going back in time one hundred years because that's what our family is doing."

Peerless is focused heavily on seven flavors of moonshine (called Lucky, because their neighbor is the Kentucky state lottery office), but they also have rye and bourbon coming, with a 2019 release date.

Taylor and his staff do a very detailed tour, including the bottling line. One nice touch: in the fermentation area, where they had six tanks in operation when I was there, you're allowed to stick your finger in and taste each one. If they offer to let you do it, don't miss it. It's enlightening to experience how corn mash changes dramatically in a few days, from a sweet slurry that tastes like creamed corn to the final sour, watery mash that goes into the still.

A couple of interesting historical tidbits on the tour: Henry Kraver's grandson, Corky Taylor's father, was an aide to Gen. George S. Patton in World War II, so the talk includes a little display of Patton's history.

And you can't miss the big "DSPKY50" stenciled all around, including on the floors: all U.S. distilleries have a DSP number—it stands for "Distilled Spirits Plant," your state-issued license to make alcohol; the lower the number, the older the license. When Peerless started in 2014, its license number was going to be 20812. Taylor's family history was so important to him that he hired a lawyer and fought to get the government to let him reinstate his great-grandfather's number, fifty.

"Fifty is a big deal to us," he says.

5   COPPER & KINGS

1121 E. Washington St., Louisville; 502-561-0267, www.copperandkings.com. Tour and tasting $15. Hours: 11 A.M.–5 P.M. Saturday–Thursday and 11 A.M.–9 P.M. Friday. Mini-tours available for $5.

Louisville's Butchertown neighborhood is exactly what it sounds like: at one time, it was where the city's abattoirs were located. Today, it's rapidly gentri-

At Copper & Kings in Louisville, the three alembic stills are named after women in Bob Dylan songs.

fying, with breweries, farm-to-table restaurants, and craft stores. Vendome, the famous still-maker, is just four blocks away.

Copper & Kings' property is eye-catching. Focused on environmentally sound practices, the front buildings are repurposed shipping containers that hold the small gift shop and the tour waiting area. There's a monarch butterfly refuge, the inspiration for their black-and-orange color scheme, and a big open area with a reclaimed-water reflecting pool and seating for hanging out at night.

The interior is just as much fun, filled with local art and a spectacular view of Louisville from a deck on the top floor. All the branding is focused on a music theme (look closely at the logo and you'll see it was inspired by old record labels, with the dot on the *i* in King's positioned to be the hole). Positioned at the front windows, the three copper stills, with graceful shapes and swirling caps, are named Sara, Magdalena, and Isis, women's names in the songs on Bob Dylan's album *Desire*.

Brandy is a "soft" alcohol, acquiring its flavors from fruit, but it can be tem-

Copper & Kings brandies include a wide variety of styles, some of it aged in a barreling room with music piped in to cause vibrations in the alcohol.

peramental, so the graceful shapes of alembic stills are designed to handle it gently, preserving flavors and aromatic qualities.

Owners Joe and Leslie Heron are beverage entrepreneurs from South Africa who started Nutrisoda and Crispin hard cider. While working with a Louisville flavor company on their Nutrisoda line, they fell in love with the city and the idea of building a distillery. Seeing America as a wide-open market, they decided to focus on brandy because of its importance in American history: before bourbon came along, brandy was big business, and there were once more than 100 brandy makers in Kentucky.

Brandy is all fruit, distilled in the same way that corn mash is distilled for whiskey. Copper & Kings is using three grape varietals—muscat alexander, chenin blanc and French columbard—all grown and turned into juice in California, as well as up to eleven varieties of apples.

Brandy has to be aged in barrels for a minimum of two years or it's classified as "immature." The barrels have to be American oak, although there's no requirement for charred or new barrels. Copper & Kings is working with that, swapping barrels with breweries, particularly stouts and porters, and using things like used Olorosa sherry, tequila, and cognac casks.

Down in the cellar aging room, you'll get a definite whiff of "angel's share," the aroma of alcohol evaporating from barrels, and you might need to cover your ears: Copper & Kings is experimenting with "resonance aging," in which music is played around resting barrels to create vibrations in the liquid, hurrying contact with the wood. There are five subwoofers and twenty-two treble speakers down there, playing lists from Spotify—with a lot of Dylan in the mix, of course.

6   LIMESTONE BRANCH DISTILLERY

1280 Veterans Memorial Hwy., Lebanon, Ky.; 270-699-9004. Hours: 10 A.M.–5 P.M. Monday–Saturday, 1–5 P.M. Sunday. Tours on the hour, with the last leaving at 5 P.M.

Some distilleries are interesting because of their method, their facility, or their equipment. Some are interesting because of the people who run them.

Limestone Branch, an easy one hour and fifteen minutes from Louisville, is the latter: tour guide Stephen Fante gives tours as loud as an auctioneer, with dramatic flourishes and a wicked sense of humor.

He starts with a talk on family history, which is considerable: the family names here are Dant and Beam, both Kentucky royalty. Figuring out the relationships between the various Beams is biblical, without quite so many "begats." It's believed that sixty-eight brands of bourbon can trace their relations to the Beam family, but Limestone is the only one still owned by Beams.

Owners Steve and Paul Beam are the great-great-grandsons of Joseph Washington Dant, who started distilling sour mash whiskey in 1836. Their paternal great-grandfather, Minor Case Beam, was the eldest son of Joseph Beam, who was the grandson of Jacob Boehm, a Kentucky pioneer who supposedly crossed the Cumberland Gap with a still strapped to his back. (Have I lost you yet?)

The current Beams opened a new distillery in 2012, making mostly moonshines and corn whiskeys under several names, including Yellowstone. The original Yellowstone came from the Dant side of the family, a straight high rye that was named for the opening of the country's first national park.

Steven Fante shows off the "barreling room," a small trailer on the grounds of Limestone Branch Distillery in Lebanon, Ky.

There's a well-stocked gift shop and a full tasting bar, as well as a rose garden and botanical garden going in outside. But the best part of the tour is Fante's explanation of a case full of family antiques, including a yeast jug, called a dona. Yeast strains were precious to whiskey families, because yeast is the difference between the flavor of your whiskey vs. your neighbor's. It had to be protected from thieves and kept cool so the yeast would multiply more slowly, cutting down on mutations. So donas were often dropped into the cistern at night for safekeeping. The one at Limestone Branch was swabbed to get samples that were used to create the yeast they're using today.

The tour includes a quick walk through the small distilling plant in the back, leading through a trailer of fifteen-gallon barrels that Fante calls "the cutest little barrel house in all of Kentucky."

7 CORSAIR DISTILLERY

400 E. Main St., #110, Bowling Green, Ky.; 270-904-2021;
www.corsairdistillery.com. 11 A.M.–5 P.M. (CST) Tuesday–Saturday.
Tours $5 with tasting or $8 with choice of souvenir glass.

8 CORSAIR DISTILLERY

601 Merritt Ave., Nashville, Tenn.; 615-200-0320. Tours $8 with tasting or $10 with
choice of a souvenir glass. Tour hours: 3:30–6:30 P.M. Tuesday–Friday, 12:30–5:30 P.M.
Saturday, 12:30–4:30 P.M. Sunday. (Also in Nashville: "Brewstillery" tours at 1200 Clinton
St., $8 with tasting, $10 with choice of souvenir glass, 3:30–6:30 P.M. Tuesday–Thursday,
2:30–5:30 P.M. Friday, 12:30–5:30 P.M. Saturday and 12:30–4:30 P.M. Sunday.)

So many distilleries play up their history and lineage. Corsair doesn't bother with that: they're among the hippest things happening in craft distilling today. You'll spot the distinctive black-and-white label of three cool guys behind bars in places like New York and Atlanta.

Yes, it is a little confusing: there are actually three Corsair Distilling facilities, one in Bowling Green and two in Nashville. How to decide which to visit? They're so different, I'd suggest visiting at least two of the three, starting with the original facility in Bowling Green, a pleasant college town not far from the Tennessee border. One thing that brings visitors to Bowling Green, besides Western Kentucky University: a lot of baby boomers come for the Corvette factory and museum, then stop in for a distillery tour. But it's just an accident of history that the distillery's original location is there.

Owners Darek Bell and Andrew Weber (Bell's wife, Amy, is the third partner) grew up in Nashville, best friends since childhood. Both studied distilling in college, but for making biofuel, projects like powering school buses on recycled cooking oil. Distilling large amounts of used cooking oil is hot, messy, and awfully smelly. So they decided to put their knowledge to work on something much more fun. They started in Bowling Green because Tennessee's laws at the time were much stricter. When the laws changed, they opened a second facility in Nashville and then added a third that's more focused on brewing.

Before they opened the distilleries, Darek Bell went to Scotland to study whiskey making. One morning at 6 A.M., he snapped a picture of three insouciant fellows walking out of a pub after a very long night. (In at least one version of the story I was told, the pub was on fire.) The picture, rendered in black and white, became their logo. Customers often ask if it's from the

Corsair's eye-catching logo—"the three random dudes"—captures the young, creative spirit of the distillery.

movies *Reservoir Dogs* or *The Blues Brothers*, but I like the name one of my tour guides gave it: "The Three Random Dudes."

That hands-in-pocket look is fitting for Corsair's product lines, which are extensive and distinctive: I counted twenty-three different bottles on one stop, and there will certainly be an entirely different lineup by the time you visit. They're doing things you won't see anywhere else, like quinoa whiskeys, triple-smoked whiskeys, barreled gins, and a rye they call Ryemageddon.

The Bowling Green facility is in an old department store on the town square, with polished wooden floors and leather sofas and chairs for hanging around. Downstairs, the distilling rooms are quirky and a lot of fun: bits and pieces of the old department store mannequins end up in odd places, like a hand sticking out of a box of bottle tops, and you'll spot crazy names and pictures painted all over the place. When I was there, they were playing with an Old Tom sweetened gin they were calling Major Tom, so an artistic employee had painted the barrel heads with images of the late David Bowie's stage personas.

Meanwhile, in Nashville, there are now two Corsair facilities. The "brewstillery," on the same street as the popular Nelson's Green Brier Distillery (see the next listing), is where Corsair is making beer using some traditional whiskey techniques, such as barrel aging, and also making whiskeys with malt.

The best location for a tour, though, is the Corsair distillery in the Wedgewood Houston neighborhood. There's a deck with outside seating, and a small tasting room with a copper bar and stools made from old barrels. Because Tennessee allows the sale of cocktails, the tour groups here tend to be younger locals. You can take a cocktail with you when you go back into the distilling facility, which makes for a relaxed atmosphere (although not quite as quirky as the Bowling Green location). They're focusing on bourbons and whiskeys here, so the tour includes an unusual 800-gallon still called Papa Smurf, with a bulbous top that concentrates alcohol similarly to the way a continuous still does.

There's nothing random about it, dudes: Corsair's distillers know exactly what they're doing.

9 NELSON'S GREEN BRIER DISTILLERY

1414 Clinton St., Nashville; 615-913-8800, www.greenbrierdistillery.com. Tour and tastings 11 A.M.–5 pm. Monday, 11 A.M.–6 P.M. Tuesday–Saturday, 11 A.M.–5 P.M. Sunday; $10, $5 with military ID. (Private tours available for $15 per person, $250 minimum weekdays, $400 minimum weekends.)

If the romance and history of distilling is what attracts you, it's hard to resist Belle Meade. The whiskey produced by Nelson's Green Brier has gained a devoted following in the South with a great story.

You'll find the distillery on a long street of brick warehouses not far from downtown Nashville. If you see a stir of people and tour buses down the block, that's the Marathon Motors building: the site of an early car manufacturing plant, it now holds the store Antique Archaeology, famous from the TV show *American Pickers*. (It also houses the "brewstillery" for Corsair Distillery.)

Nelson's story starts in Germany in the mid-nineteenth century. The Nelson family decided to immigrate to America with their six sons, so they sold everything and converted their assets into gold bars, sewn into a specially made suit worn by the family patriarch. The family got passage on the *Helena Sloman*, the same wind-and-steam sailing ship that had already brought the Steinway and Heinz families. The Nelson family wasn't so lucky, though. The ship ran into a storm and many passengers went overboard, including the senior Nelson, who was pulled down by the weight of the gold in his suit and drowned.

The rest of the family reached America penniless. The oldest son, fifteen-year-old Charles, went to work at a number of jobs, finally reaching Cincinnati as a butcher. The pigs that came to him had usually been fattened on corn mash from distilleries, drawing his attention to the money that could be made selling whiskey. He moved to Nashville and opened a store selling meat, coffee, and whiskey. (His coffee was sold at the Maxwell House Hotel, where it was given the moniker "Good to the last drop.")

Nelson sold so much whiskey that he built his own distillery in Greenbrier, Tennessee, in 1870, making classic whiskey filtered through sugar maple charcoal. By 1885, he was making 2 million bottles a year, one of the state's most successful operations.

Charles died in 1891, but his wife, Louisa, took over and ran the distillery until Tennessee enacted its own Prohibition in 1909, several years before the rest of the country. Realizing it was going to be a long dry spell, Louisa Nelson sold the remaining stock and shut down.

That's where it would have ended, if not for brothers Andy and Charlie Nelson, the Nelsons' great-great-great-grandsons. Both in college studying philosophy, they were giving a party and wanted meat for a barbecue. Their dad took them out to the country to a butcher he knew. While they were stopped for gas, they spotted a historical marker noting the nearby location of the original Nelson distillery. The butcher, it turned out, was located across from the only remaining distillery building, an abandoned warehouse.

The family had heard stories about their whiskey-making ancestors but knew little about it. Andy and Charlie got so interested, they visited the local historical society and discovered their history, along with two partially empty bottles. They ended up changing their majors to business and eventually opened a distillery, making whiskey inspired by samples of the original.

Today, they operate a beautiful distillery crafted from wood and brick, with a sizeable gift shop, historical displays—including a model of that sailing ship—and a tour that includes their hybrid pot still, named Louisa. They also make a "clear" whiskey (aka moonshine), but it's Belle Meade that has put them on the map, with distribution in sixteen states and the District of Columbia.

Maple-filtered in the Tennessee style, it's a rye-based whiskey aged six to eight years in deep-char barrels; they're also making variations aged in sherry and cognac barrels.

The tour is worth a stop for the history, a good explanation of that Tennessee style, and generous pours at the polished wood tasting bar.

10   TENN SOUTH DISTILLERY

1800 Abernathy Rd., Lynnville, Tenn.; 931-527-0027, www.tennsouthdistillery.com.
Tour and tastings 9 A.M.–5 P.M. Monday–Saturday.

If it's a nice day for a drive, you'll find a pretty one in the trip to Giles County, past the steep valleys of western Tennessee and through the picturesque town of Lynnville. Cindy and Clayton Cutler, their brother-in-law Blair Butler, and their son-in-law Colin McLaughlin have built a small distillery in a field reached by curving country roads.

Clayton Cutler was in the ink-jet refilling industry, traveling the world but also seeing the end of ink-based printing on the horizon. He decided it was time to do something completely different, so he teamed with his brother-in-law, a radiologist, to try their hand at Tennessee whiskey. Colin became the distiller, while Cindy, who had spent her life in the wedding industry, making cakes and dresses, joined in to run the gift shop and tasting room.

The distillery is built on twenty-eight acres in Giles County, using corn grown nearby by a farmer who also grows for Martha White and White Lily. The building is small, but they're managing to turn out a barrel a day four times a week, for 25,000 cases a year: All Mighty Shine, Big Machine Platinum Vodka, and Clayton James Tennessee Whiskey.

11   PRICHARD'S DISTILLERY

11 Kelso Smithland Road, Kelso, Tenn.; 931-433-5454, prichardsdistillery.com.
Tours 9 A.M.–4 P.M. Monday–Friday, 9 A.M.–3 P.M. Saturday. (Also Prichard's
at Fontanel, 4105 Whites Creek Pike, Nashville; 615-454-5991.)

If you can't resist a stop in Lynchburg for Jack Daniel's, include time for a stop at this nearby distillery, built in what used to be a schoolhouse. (There's a second Prichard's location in Nashville, on the grounds of Fontanel, the historic home once owned by country singer Barbara Mandrell, but the majority of the distilling happens in Kelso, about fifteen miles from Lynchburg.)

Phil Prichard's obsession is colonial-style rum. Originally a dental technician, he made it as a hobby until friends urged him to go pro.

The Prichards picked Lincoln County because the location near the George Dickel and Jack Daniel's distilleries meant it was one of only three counties that allowed distilling before the laws changed in 2009. But the location also turned out to have an old school house, built in 1939 and used until 1979. It had been turned into a community center for the little farming community, as well as the location of the volunteer fire department. The fire department built a tall addition to house a fire engine, but then couldn't raise the money to buy one. So there was a tall space on one side that was perfect for a still.

Using table molasses, which is sweeter than blackstrap molasses, they're making rum and aging it for up to twelve years, and they've added seven whiskeys, including a bourbon, a rye, and a corn-based Tennessee whiskey made with white corn that has a distinctive, cornbread aroma.

One of their biggest hits, though, is a cream liqueur, Sweet Lucy Cream, flavored with orange, apricot, and bourbon flavoring with a base of stabilized cream. (There's also a spicy version of the basic "Sweet Lucy" liqueur, called Sweet Lucifer.)

Lucy was born in a duck blind: hunting was a family activity, and Prichard's father used to soak fruit in whiskey and take it along, to take the chill off while they were shooting ducks on the water. After the hunt, everyone got to take a sip to warm up. The Prichards were a hunting and drinking family, but not a cursing one, so when they felt strongly about something, they would declare, "Sweet Lucy!"

TRAVELER'S NOTE: If you go on into Lynchburg to visit Jack Daniel's, make your plans in advance. It's a famously dry county and a very small town. Hotel choices are limited, and while Miss Mary Bobo's Boarding House is renowned for southern cooking, reservations can be tight and it closes at 4:30 P.M. If you get there after 6, you may find the downtown locked up tight.

12 KNOX WHISKEY WORKS

516 W. Jackson St., Knoxville, Tenn.; 865-525-2372, www.knoxwhiskeyworks.com.
Tours: hourly from 5–10 P.M. Thursday and Friday and 2–10 P.M. Saturday.
Tour and four-flight tasting, $10. Cocktails also available.

Knoxville's downtown area, called Old City, has undergone a renaissance in the last decade, with the downtown Market Square revitalized and the blocks around it springing up with farm-to-table restaurants and craft breweries. With both the University of Tennessee and the headquarters of Scripps Broadcasting, home of the Travel Channel and HGTV, there's a young and energetic vibe to the food scene here.

On the edge of the district, overlooking the railroad yards, West Jackson is a steep street lined with brick warehouses that were almost abandoned until recently. But new craft businesses are taking over the old buildings. Just up the street from Knox Whiskey Works, another distillery, PostModern Spirits, wasn't yet open when I visited in spring 2017, but will eventually make this an even richer stop for craft distilling fans.

One distinction of Knox Whiskey Works is an almost all-female crew: head distiller Miranda White is only in her mid-twenties, but she's already a veteran who worked at Popcorn Sutton, a moonshine distillery, before taking the reins at Knox. And five of the nine co-owners are women.

When I visited, they were sourcing neutral grain spirits from Indiana for their vodkas and gins, but they're making their own whiskey from Hickory Cane corn, an heirloom variety grown in eastern Tennessee, that's just starting to be released from barrels, starting with five-gallon vessels and working up to larger barrels with longer aging. They are doing things with those neutral grain spirits, though: Tailgate Vodka is flavored with blood oranges (orange for the University of Tennessee, of course); a lightly sweetened vodka; and two excellent gins, an unusual pink gin aged in Napa Valley cabernet barrels and a New West gin, Jackson Avenue, that's found a following in local craft cocktail bars. They also offer two floral/herbal liqueurs: Fleur De Vie, for use in cocktails, and a caffeinated coffee liqueur, using a cold-brew infusion from Counter Culture coffee beans, that's like espresso with a kick.

The distillery itself is a small space, but it's worth a look for the unusual still: handmade from a dairy chilling tank and topped with copper, they found it on Craig's List after a distillery in Nashville shut down.

"Keeping it classy" is what my tour guide, a very energetic graduate student named Ashley Shepperd, told me with a grin.

Oscar Getz: The Sweetest Little Whiskey Museum in Kentucky

In Kentucky, one town is so associated with making bourbon that the name itself has almost become a trademark for it: Bardstown. Less than an hour's drive southeast of Louisville, it's a busy little place, with a picturesque and strollable downtown packed with art, antiques, crafts, and all things bourbon-related.

While I didn't highlight a Bardstown craft-distillery stop, Willett Distillery would be a great choice. Or you can vary your experience with a larger, more mainstream facility such as Heaven Hill or Barton 1792.

Whatever you do, though, there's one stop that any bourbon fan ought to make: the Oscar Getz Museum of Whiskey History, just a few blocks from the town center.

It's located on the first floor of Spalding Hall, 114 North Fifth Street. Built in 1826, the building has served several roles: it was originally the St. Joseph College and Seminary, then a Civil War hospital, a boys' orphanage and finally a boys' prep school before finally housing the whiskey museum and the Bardstown Historical Museum.

It's a wonderfully creaky and charming old place, with wood floors, elaborate fanlights over some of the doors, and room after room of whiskey history, artifacts, and memorabilia covering the earliest processes of distilling to

In downtown Bardstown, you'll find the Oscar Getz Museum of Whiskey History on the first floor of a nineteenth-century building that used to be a boys' school.

the post-Prohibition era. It's not huge, but you can spend a good ninety minutes there, and if you're a buff, it really will be a good ninety minutes.

The front-office staff is friendly and helpful if you need advice on where to eat lunch or any Bardstown trivia. You also sometimes find old bottles and decanters for sale.

The admission is by donation, and the hours are 10 A.M.–4 P.M. Tuesday–Saturday, noon–4 P.M. Sunday (closed Monday) from Nov. 1 to April 30, and 10 A.M.–5 P.M. Monday–Friday, 10 A.M.–4 P.M. Saturday and noon–4 P.M. Sunday from May 1 to Oct. 31. Details: 502-348-2999, or www.whiskeymuseum.com.

Other Craft Distilleries in Kentucky and Tennessee

Bluegrass Distillers. 501 W. Sixth St., Lexington; 859-253-4490, bluegrass distillers.com. Tour hours: 11 A.M.–5 P.M. Monday–Thursday and 10 A.M.– 5 P.M. Friday–Sunday.

Boone County Distilling Co. 10601 Toebben Dr., Independence; 859-282-6545, www.boonedistilling.com. Hours: noon–5 P.M. Wednesday–Saturday, noon–4 P.M. Sunday. Tours every day at 12:30, 2:00, and 3:30 P.M.

Casey Jones Distillery. 2813 Witty Ln., Hopkinsville; 270-839-9987, www.caseyjonesdistillery.com. 10 A.M.–6 P.M. Friday–Saturday; free tours and tastings.

Castle & Key Distillery. 4445 McCracken Pike, Frankfort; 859-873-2481, www.castleandkey.com. Not yet open to the public when this book was being researched but worth keeping an eye on. It's on the grounds of the historic Old Taylor Distillery, in a building that actually does look like a castle. It was projected to return to operation in 2017 after extensive restoration.

Hartfield & Co. Distillery (formerly The Gentleman Distillery). 108 E. Fourth St., Paris; 859-559-3494, www.hartfieldandcompany.com. 9 A.M.–5 P.M. Tuesday–Friday, 10 A.M.–5 P.M. Saturday. Offers shorter "City" tours Tuesday–Saturday and the longer "County" tour at 2 and 4 P.M. Saturdays. (Prichard and Bail Cocktail Bar located at the same address.)

Kentucky Artisan Distillery. 6230 Old La Grange Rd., Crestwood; 502-822-3042, www.kentuckyartisandistillery.com. Hours: 10 A.M.– 4 P.M. Tuesday–Friday, 10 A.M.–2 P.M. Saturday. Includes Jefferson's Bourbons, Highspire Whiskey (rye), and Whiskey Row Whiskey Bourbon. Tours: $10; $5 for active military.

MB Roland Distillery. 137 Barkers Mill Rd., Pembroke; 270-640-7744. Hours: 10 A.M.–6 P.M. Monday–Thursday, 9 A.M.–6 P.M. Friday and Saturday, 1–6 P.M. Sunday. Tours: $5 with shot glass, or $25 for a special distiller's tour with a behind-the-scenes tasting and rocks glass.

Second Sight Spirits. 301B Elm St., Ludlow; 702-510-6075, secondsightspirits .com. Hours: Noon–5 P.M. Thursday, noon–8 P.M. Friday and Saturday. Tours at 12:30, 2, 4 and 6 P.M. (no 6 P.M. tour on Thursdays). Free tour and tasting.

Silent Brigade Distillery. 426 Broadway St., Paducah; 270-709-3242, www.silentbrigadedistillery.com. Hours: 11 A.M.–1 A.M. Tuesday–Thursday, 11 A.M.–3 A.M. Friday and Saturday, 1–10 P.M. Sunday.

Town Branch Distilling. 401 Cross St., Lexington; 859-255-2337, www.kentuckyale.com. Part of Alltech Lexington Brewing & Distilling Co., which makes Kentucky Bourbon Barrel Ale, aged in used bourbon barrels, this is one of the few brewery/distillery combinations. Variety of tours, starting at $15 and including a $30 master-distiller tour and a brewer/distiller tour for $50.

Wilderness Trail Distillery. 4095 Lebanon Rd., Danville; 859-402-8707, www.wildernesstracedistillery.com. Hours: 10 A.M.–4 P.M. Tuesday–Saturday (closed 12–1 P.M. for lunch). Includes the first craft-distillery rickhouse (barrel-aging facility) in the state.

Willett Distillery. 1869 Loretto Rd., Bardstown; 502-348-0899) www.kentucky bourbonwhiskey.com. Hours: 9:30 A.M.–5:30 P.M. Monday–Saturday (year-round), noon–4:30 P.M. Sunday (April–December). A distillery with long history, it was restored and returned to distilling in 2012 in Bardstown. Tours: $12 with a tasting and Glencairn glass (military discounts available).

TENNESSEE

Cocke County Moonshine Distillery. 337 Old Knoxville Hwy., Newport; 423-248-3070, www.cockecountymoonshinedistillery.com. Free tastings.

H Clark Distillery. 1557 Thompson's Station Rd. W., Thompson's Station; 615-478-2191, hclarkdistillery.com. Tours are free; tastings are $10. Hours: 11 A.M.–4 P.M. Monday–Friday, noon–4 P.M. Saturday.

Jug Creek Distillery. 1049 Oregon Rd., Lascassas; 615-273-5847, www.jugcreek distillery.com. Tours and tastings are $10; 11 a.m.–7 p.m. Thursday, noon–8 P.M. Friday and Saturday, 1–6 P.M. Sunday.

Nashville Craft. 514 Hagan St., Nashville; 615-457-3036, www.nashvillecraft .com. In the Wedgewood-Houston neighborhood near Corsair, with local-ingredient whiskeys and gins, plus Naked Biscuit sorghum syrup. Hours: 10 A.M.–6 P.M. Tuesday–Saturday, noon–6 P.M. Sunday. Closed Mondays.

Old Dominick Distillery. 305 S. Front St., Memphis; 901-260-1250, olddominick.com. A new distillery from an old beer-distribution family, it's making vodkas and whiskey. Tours: noon–7 P.M. Thursday, noon–8 P.M. Friday and Saturday, 11 A.M.–5 P.M. Sunday. $12 for tour and tasting, includes souvenir guidebook.

Old Forge Distillery. 170 Old Mill Ave., Pigeon Forge; 865-774-4126, oldforgedistillery.com. 10 A.M.–9 P.M. Monday–Saturday, noon–6 P.M. Sunday. Free tastings; tours by request.

Old Glory Distilling Co. 451 Alfred Thun Rd, Clarksville; 931-919-2522, oldglorydistilling.com. White rum, vodka, moonshine, and corn whiskey. Tasting bar open daily from 11 A.M.–6 P.M. Monday–Saturday and 1–6 P.M. Sunday; tours $7, offered hourly on Saturdays until 5 P.M.

Ole Smoky. 903 Parkway, Gatlinburg, and 131 The Island Drive, Pigeon Forge; no phone number available, olesmoky.com. Moonshines, flavored moonshines, whiskeys, and flavored whiskeys. 10 A.M.–11 P.M. daily in Pigeon Forge; 10 A.M.–10 P.M. Monday–Saturday and 10 A.M.–10 P.M. Sundays (moonshine sales only noon–7 P.M.) in Gatlinburg.

Popcorn Sutton Distillery. 830 U.S. 25 West, Newport; 423-532-8501, popcornsutton.com. Focused on the legacy of the historic Tennessee moonshiner. Free tours 9 A.M.–3 P.M. Monday–Friday; call for reservations (required).

PostModern Spirits. 205 W. Jackson Ave., Knoxville; www.postmodernspirits .com. Focuses on American single-malt whiskey and barrel-aged gin. Tours: 6 P.M. Thursday, 2 and 4 P.M. Saturday, 2 P.M. Sunday.

Pyramid Vodka. 802 Royal Ave., Memphis; 901-576-8844, www.pyramid vodka.com. Vodka produced by a restaurant family in the old Firestone neighborhood near downtown Memphis. Tours 2–4 P.M. Wednesday–Friday, 10 A.M.–1 P.M. Saturdays.

Short Mountain Distillery. 8280 Short Mountain Rd., Woodbury; 615-216-0830, www.shortmountaindistillery.com. Tours include organic farming practices and a trail walk. Includes a café on site. Tours are $10, including a shot glass, 11 A.M.–4 P.M. Friday–Sunday. (There's a moonshine cocktail class on Saturdays at 11:30 A.M. for an additional $15.)

Southern Pride Distillery. 108 Smith Mill Rd, Fayetteville; 931-433-9137, www.southernpridedistillery.com. Tasting room with free tours; no appointment required. Hours: 8 A.M.–4 P.M. Monday–Friday, 8 A.M.– 2 P.M. Saturday.

Sugarlands Distilling Company. 805 Parkway, Gatlinburg; 865-325-1355, www.sugarlandsdistilling.com. Tennessee-style sour mash moonshine. Hours: 10 A.M.–10 P.M. Monday–Thursday, 10 A.M.–10:30 P.M. Friday and Saturday, and noon–6:30 P.M. Sunday (summer hours 10 A.M.–11 P.M. Monday–Saturday, noon–7 P.M. Sunday). In addition to free daily tours, it also offers blended sampling tours ($15), distiller workshops ($40), and "Distiller for a Day" workshops ($250).

Recipes

BUTCHERTOWN PUNCH

From Copper & Kings, in Louisville's Butchertown district.

MAKES 7 (6-OUNCE) SERVINGS

3 lemons

3 ounces sugar

8 ounces brandy

2 ounces peach liqueur

2 ounces dark rum

4 ounces brewed black tea

3 lemons, zested and juiced

16-ounce block of ice, preferably made from filtered water

Peel the zest from the lemons, scrapping away any bitter pith. Place the peels and the sugar in a large bowl, rubbing or mashing to release the citrus oils. Let stand for 30 minutes to an hour.

Juice the peeled lemons and add to the sugar and lemon peels, stirring until the sugar is dissolved. Strain to remove the lemon peels.

Add the brandy, peach liqueur, rum, and black tea. Stir to mix well. Add a 16-ounce block of ice and let stand 30 minutes, so some of the ice melts to dilute the punch.

Serve garnished with lemon wheels.

KENTUCKY AND TENNESSEE 125

THE REVOLVER

From Brian Lorusso, beverage manager for Dogwood Southern Table & Bar in Charlotte, North Carolina. He combines Tennessee whiskey with Damn Fine coffee liqueur from Durham Distillery and orange and fig bitters from Crude in Asheville, North Carolina. You could substitute other orange bitters, such as Reagan's or Fee Brothers, and another coffee liqueur.

MAKES 1 DRINK

1 ½ ounces Belle Meade Whiskey

½ ounce Damn Fine coffee liqueur

Strip of orange peel

3 dashes orange bitters (preferably Crude orange and fig from Asheville, N.C.)

Stir all the ingredients together with ice in a mixing glass. Strain and serve straight up in a chilled glass, garnished with the orange peel.

A SOUTHERN MANHATTAN

From Brian Lorusso of Dogwood Southern Table & Bar in Charlotte, North Carolina. Lorusso's version of this classic is truly a three-state creation — Belle Meade from Tennessee, Southern Amaro from High Wire in South Carolina, and coffee and cocoa bitters from Crude in Asheville, North Carolina. You could use any good whiskey and amaro, of course.

MAKES 1 DRINK

2 ounces Belle Meade Whiskey

1 ounce High Wire Southern Amaro

3 dashes Crude coffee and cocoa bitters

Luxardo or Jack Rudy cherry for garnish

Stir all the ingredients in a mixing glass with ice. Strain into a chilled glass and garnish with a cherry.

Bitter Pill: Prohibition and the Medicinal License

The passage of Prohibition nationwide in 1919 obviously had earth-shaking ramifications for the American distilling industry. Besides forcing liquor underground and into the hands of organized crime, it was a major blow to the business and craft of distilling. And while it was made possible by the passage of the Eighteenth Amendment to the Constitution, it went on even longer in some places. As many as nineteen states, many in the South, had passed their own anti-liquor laws by 1916, some as early as 1905.

Since the repeal amendment left the regulation of alcohol production and sales in the hands of the states, the legacy lingered in many places, with dry counties that still exist and cities that took years to return to liquor by the drink. Mississippi didn't repeal statewide Prohibition until 1966.

One angle of Prohibition comes up over and over in the story of craft distilleries that are reigniting family histories, particularly in Kentucky: the medicinal license.

When Prohibition passed, a loophole allowed the granting of a handful of special permits to provide alcohol for medical reasons. A half-dozen distilleries, including Stitzel, Glenmore, Schenley, Brown-Forman, National, and Frankfort, were granted licenses that allowed them to store existing whiskey and sell it to pharmacists, who could dispense it with a doctor's prescription. If you could convince your doctor that you suffered from a variety of ailments, from "women's complaint" to headaches, nerves, or sleeplessness, you could get a prescription for a pint of liquor.

Before Prohibition took effect on January 17, 1920, a number of distilleries sold whatever stock they had to the distilleries that got the permits. Toward the end of Prohibition in 1929, the U.S. government also allowed occasional "distilling holidays" for making a little more to refresh the "medicinal" supplies. Still, there was very little new whiskey made (legally, anyway) for more than thirteen years.

Even with those two loopholes, the effect on the distillery business was devastating. Many distillers simply closed down, never to reopen. And since good whiskey takes years to make and age, supplies at those that stayed in business with medicinal permits were often wiped out. When the Volstead Act was repealed, it happened in the early years of that other American disaster, the Great Depression. Many distilleries couldn't afford to start from scratch making something that would take years of capital investment before there was any profit.

Although some famous names managed to restart or were created in the 1930s, the landscape of American distilling was changed for decades to come.

Georgia, Alabama, and Florida

LIQUOR TRAIL

5

AS YOU MOVE INTO the middle South, the number of distilleries drops. In the area sometimes dubbed "the buckle of the Bible Belt," state governments have been slower to adopt changes in laws, and local attitudes have historically been disapproving of alcohol. But the lure of tax revenue and the economic success of craft brewing are beginning to change the climate for craft distilling. Georgia and Florida in particular have eased some restrictions, allowing growth.

When Georgia modernized its laws in 2015 to enable craft breweries to sell on site, it included an unusual provision for distilleries, allowing visitors to include a "souvenir" bottle with the cost of the tour. So tours are priced accordingly, usually around $30 to $50, depending on the price of a bottle. (Most also allow you to pay a much smaller price for a tour and tasting only, typically $5 to $7.)

Florida is even more welcoming, with a major overhaul of the state's Prohibition-era regulations passed in 2013 that has brought an explosion of distilleries and tasting rooms. The new law included a microdistillery category for operations that make less than 75,000 gallons a year and a limited amount of on-site sales to visitors. You're allowed to buy two bottles a year from each distillery, but sales are based on brands. So if a distillery creates multiple brands, each with a separate SKU (stock-keeping unit, used in inventories), you can potentially buy a whole lot more than two bottles per place.

With the state's rich agricultural scene, particularly in tropical fruits, and with so much interest in locally produced products, small Florida distilleries are able to stretch out and get creative. You can find anything from blueberry vodka to rhum agricole from the state's huge sugarcane crop. One distillery, Fish Hawk, is way off the beaten path in the woods near Ocala, but it was making a wild assortment of things when I was there, including tangerine-based brandy and Florida-sourced red absinthe.

Alabama is changing more slowly, with just a handful of distilleries state-

Mark Allen transfers freshly made corn whiskey into barrels one bucket at a time at Lazy Guy Distillery in Kennesaw, Ga.

wide. With the manufacture of alcohol illegal since 1915, when the state enacted early Prohibition, the state modernized laws in 2011 to allow breweries to open. That also allowed distilleries to begin to operate with tasting rooms that can bring important income to build a business.

You don't have to be big to be good, though: John Emerald, the one distillery that was open for tours and tastings when I visited, is producing a single-malt that has been winning awards and acclaim. Others were on track to open within the next few years all over the state.

This tour will start in north Georgia and head south, down the west coast of Florida and then back up, ending in St. Augustine.

2950 Moon Station Rd., Kennesaw, Ga.; 770–485-0081, www.lazyguydistillery.com.
Tours: Noon to 6 P.M. Saturdays (noon to 5 P.M. in winter); $5 or $35 including
a bottle (higher for some specialty bottles). Private tours are available
on other days for $35 if you arrange in advance.

You can't avoid history in Kennesaw: the Civil War may have ended more than 150 years ago, but that great conflict left scars over this part of Georgia that are still a part of its legacy.

At Lazy Guy Distilling, where owner Mark Allen and his assistant, Cody Chinn, are anything but lazy, your visit starts with history. The office and tasting room are in a house built in 1885, just a couple of blocks from the small downtown, beside railroad tracks that were the start of the Great Locomotive Chase: Union soldiers commandeered a Confederate locomotive, The General, that was refueling at the little settlement of Big Shanty, later renamed Kennesaw, and took off for Northern-held Chattanooga. (The distillery is a couple of blocks from the Southern Museum of the Civil War and Locomotive, if you want to learn more.)

The wooden barn behind the house, which holds the distillery, is believed to have been built well before the war, in 1830, which makes it one of the oldest wooden structures in the state. Gen. William T. Sherman burned almost everything made of wood during his march across Georgia. (At Lazy Guy, they claim Sherman may have spared the barn because it had a still in it already, but there's no proof of that.)

The name Lazy Guy is a joke, of course: Allen owns an internet technology consulting company, and the distillery is his "retirement plan." But nothing about having two men running a distillery while working full-time is retiring or lazy.

Chinn, twenty-seven when I met him, is what you might call an organically grown distiller. A native of Rosine, Kentucky, his family was friends with Charles Medley of the Medley Brothers, but Chinn says he actually learned distilling from Amish people who lived around Rosine: they made their own spirits for fuel and for drinking, allowed in some Amish communities, particularly among young men.

In high school, Chinn used to smuggle his homemade whiskey to school in the wheel of his car, selling it on the side to friends. At Bowling Green State University, he made wine and sold it to sorority girls to pay for his tuition. After college, he was working in Web design when he came by Lazy Guy on a tour. He liked what Allen was doing so much, he offered to work on weekends in exchange for free liquor.

The two of them manage to make 4,500 bottles a week, including several kinds of bourbon, corn whiskey, rye, and a liqueur called Snow Cream. Since they work on the weekends, you stand a good chance of seeing something actually happening if you visit on a Saturday, like the back-breaking work of loading newly distilled alcohol into barrels one bucket at a time.

The barn itself is a beauty for woodworking fans: made of rough-hewn logs and hand-planed boards and beams, some of the original mortar is still holding the logs together. You'll see the hybrid pot-and-column still, a style that's popular all over Georgia, and get a stroll through a packed barrel-aging room.

So many people his age are making beer that I asked Chinn why he's so drawn to making whiskey instead.

"Brewing is a science," he said. "You control all the variables. Distilling is an art. You adjust as you go." That makes distilling endlessly fascinating: until it comes out of the barrel, you don't really know what you're going to get.

2   ASW DISTILLERY

199 Armour Dr. NE, Suite C, Atlanta, Ga.; 404-590-2279, www.aswdistillery.com.
Tours: 4:30–7:00 P.M. Thursday and Friday, 2–5 P.M. Saturday.
Tour prices range from $8 to $60, depending on your "souvenir" bottle.

Jim and Kelly Chasteen and their co-owner, Charlie Thompson, are going all out in style. Even though their distillery is located in a small area of industrial buildings near downtown Atlanta (SweetWater Brewing is a neighbor), the inside of their distillery is all gleaming wood floors with a well-stocked gift shop area and a large tasting bar. (There's even a "white dog" loping around, named Barley.)

Kelly Chasteen showed me around with the pride of a designer-home owner: the private area hidden behind a frosted glass door labeled "Research & Development" is actually a hospitality room for private events, with a fireplace, ping-pong table and vintage wooden lockers (an homage to Atlanta "locker clubs" where people stashed "medicinal" whiskey during Prohibition). One wall is covered with a massive photographic mural of men celebrating the end of Prohibition in Atlanta's Marietta Square.

Kelly Chasteen had to search awhile to find a historic picture that showed a happy scene from Prohibition in Georgia for the wall of a private lounge for investors at ASW Distillery in Atlanta: "Most pictures of them show stills being chopped up."

"It's hard to find joyful pictures of Prohibition," she said. "Most of them show stills being chopped up."

Having a hospitality room for their seventy-five investors was important, because they want everyone to feel like they have a stake, she says. Jim Chasteen was in real estate investment trusts and traveled a lot. On weekends, they'd invite friends over so he could spend more time at home with their two kids.

"Jim was ready to come off the road," she says. He and Thompson, his college roommate, were drinking rye at one gathering when the idea came up: "How hard can it be to make whiskey?"

They hit Google, found equipment, and started making whiskey at home for friends. Finally, they bought a book, *The Business of Spirits*. Since Charlie is "a recovering lawyer," he navigated the regulations to get a license.

"The idea was to build the brand so building the distillery would be less risky," Jim says. With seventy-five investors, no one had to gamble too much money, and they became the sales force—they all ask for it wherever they go, building buzz.

ASW started with spirits that were made at a distillery in Charleston, buying wheated-bourbon barrels from MGP in Indiana and putting their spirits into them. (They call it Fiddler, because "we didn't make it here, but we fiddled with it.") Using a pair of matched copper pot stills, they're now making their own whiskeys, using a Scottish-style "grain-in" method, leaving the grain in the still instead of just piping in the fermented low wines. Their corn whiskey aims to be a bridge between vodka and moonshine, with a little corn sweetness, and they're aging apple brandy made from North Georgia apples.

Since Georgia requires a "souvenir bottle" tour price, they're playful with it, offering seven versions based on what you want to try or buy. One option includes your choice of a bag of Zapp's potato chips, so your tour really can be "all that and a bag of chips."

3   OLD FOURTH DISTILLERY

487 Edgewood Ave. SE, Atlanta, Ga.; 844-653-3687, o4d.com. Tour times: 5–9 P.M. Thursday and Friday, noon–9 P.M. Saturday. Tour and tasting $10; $30 with a bottle.

Atlanta's Old Fourth Ward neighborhood is in downtown, not far from the Sweet Auburn area that was the home of Dr. Martin Luther King. Today, it's gentrifying quickly, with $900,000 houses and plenty of joggers and baby strollers on a Saturday morning, along with restaurants and art galleries in the old buildings.

Old Fourth Distillery is growing with the neighborhood: currently located in a brick building, they've just added a bigger warehouse up the street to keep up with the demand. The original location will stay, though. Designed with a late-nineteenth-century feel, the marble tasting bar, a blackboard scribbled with specials and the rescued wood on the walls all came from a boys' school that was being torn down.

Don't miss the shelves of rare stoneware jugs stamped with the name R.M. Rose Distillery. Rose was a major figure in Atlanta history. The last legal distiller in the city before Prohibition, he's buried in historic Oakland Cemetery, where the owners like to pick juniper berries for their gin. When it opened in 2014, Old Fourth became the first legal distillery in the city since 1906, and so it's become a repository for Atlanta liquor history.

The owners are Gabe Pilato and brothers Craig and Jeff Moore. They all

Old Fourth Distillery is building on the city's whiskey-making history and industrial legacy at its facility in downtown Atlanta.

grew up in Atlanta, with the Moores going into technology and Pilato becoming a sommelier. Pilato was running restaurants in Charleston, feeling burned out on the pretensions of the wine business, when the Moores called him: "We own a building. You want to do something?"

"I'm kind of anti-snobbery," he says. "My whole thing as a somm was to find cheap wines that would blow you away." So the idea of learning another form of alcohol, one he could make himself, was enticing. Jeff Moore went to Cornell University for a quick class in distilling, and they taught themselves the rest.

"Trial and error, man," Pilato says.

Using local water—legend has it that Atlanta has particularly good water because of its history with Coca-Cola—and ingredients they source from around the South, like a fifth-generation sugarcane farm in Louisiana, they're making vodka, gin flavored with a long list of botanicals, and bourbon. They've also added a not-too-sweet gin and lemonade beverage called

Lawn Dart. The bourbon is aging in full-size fifty-five-gallon barrels and won't be released until 2019, although you could sample it at the tasting bar when I was there.

Pilato and his partners are focusing on building a business centered in Atlanta, rather than trying to become a national brand.

"We're trying to follow the SweetWater [Brewing] model. There's seven million people here [in Atlanta]. There's plenty of people to sustain it."

4 JOHN EMERALD DISTILLING COMPANY

706 N. Railroad Ave., Opelika, Ala.; 334-737-5353, www.johnemeralddistilling.com. Tours 5–9 P.M. Thursday–Friday and 3–9 P.M. Saturday.

Opelika, Alabama, near Auburn, is so close to Georgia that it's an easy detour from Atlanta. Here's hoping actress Amy Schumer finds her way there: the John Emerald tasting bar includes a board labeled BADLAD: Buy a Drink, Leave a Drink. You can buy a drink for someone and their name goes on the board in case they ever come in. Schumer has quite a night waiting if she ever stops in Opelika.

As one of the first distilleries to navigate Alabama's byzantine liquor rules, the father-son team of John and Jimmy Sharp has built a laid-back distillery in an old cotton warehouse downtown, a few blocks from the county courthouse near the railroad tracks. Like a lot of small farm towns, Opelika is trying to reinvent itself as a place that welcomes entrepreneurs, and a local-food scene is beginning to bubble up.

The Sharps were a military family, moving around a lot when Jimmy Sharp was a boy. He went into custom carpentry, a job that also involves a lot of travel to building sites. When his daughter was born, they both decided to find a family business that would let Jimmy stay put.

John Sharp's father, John Emerald Sharp, came from Scotland, and Jimmy had gotten interested in learning the family history.

"You can't learn about Scotland without whiskey," he says. He ended up going to Scotland to intern at a distillery, then came back to open one with his father, specializing in single-malt whiskey from 100 percent barley, 18 percent of it smoked over peach and pecan wood. Aged in charred oak, it's sort of a cross between bourbon and Scotch. They're also making spiced rum and gin.

They're trying to use as many Alabama products as possible, including brandy from local muscadine grapes.

"We may have found muscadine's true purpose," Sharp says. "It doesn't make good wine, but it makes good brandy."

Because it's small, the Sharp distillery can be experimental, figuring out what works. While Jimmy disdains adding wood chips to barrels to increase exposure to wood, he's figured out an unusual speeded-up barreling process that involves rapidly changing the temperature inside his fifteen-gallon barrels using a restaurant-size walk-in refrigeration unit.

"With wood chips, you're missing half the process," he says. You get some of the flavors extracted from the wood, he says, but you don't get the color and real flavor changes that come from the pressure caused by temperature fluctuations when barrels sit through summers and winters.

"To achieve [flavor] esters, you need pressure changes driven by temperature."

By putting barrels into the refrigeration unit and adjusting the temperature while opening and closing the door, he can repeatedly raise and lower the temperature inside each barrel by thirteen degrees.

"That's enough to make it swell and contract," he says.

Sharp first tried it on a single barrel and was stunned to discover the difference it made.

"In six weeks, the difference was unbelievable," he says. "It was a palpable difference. Some of it was putting two and two together. I figured out it was working and then figured out why."

5   RICHLAND RUM

333 E. Broad St., Richland, Ga.; 229-887-3537, www.richlandrum.com.
Tours 10 A.M.–4 P.M. Monday–Friday, Saturday by appointment 11 A.M.–4 P.M.
Tour and tasting price varies from free to higher with a bottle.

Can a distillery save a dying town? In Richland, Georgia, a Dutch couple is giving it a try.

Says Karin Vonk: "If we don't do it, who will?"

Richland, a two-hour drive from Atlanta's Hartsfield–Jackson International Airport, used to be a thriving farm town. In the 1960s and 1970s, the population was as high as 12,000, with a railroad station, four car dealerships, and a small downtown of shops and businesses in late-nineteenth-century buildings. Then mechanization changed agriculture, eliminating field jobs, and the town shriveled to less than 1,000 people. The train station

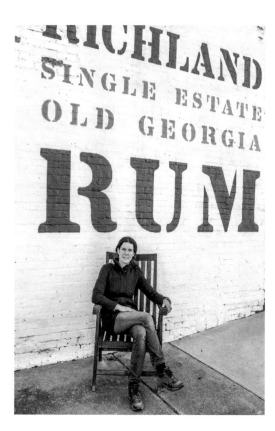

Karin Vonk's dream ("hope, dream, plan") is to use Richland Rum as a base to make the tiny town of Richland a destination for food tourism.

closed and even the tracks were dug up. The downtown buildings were shuttered and abandoned.

"It's the rural American story," Vonk says.

Vonk and her husband, Erik, lived in Atlanta for years, where Erik ran a successful staffing services company. He was born in Holland but raised in Atlanta. Karin came over in the 1970s. She was a farmer's daughter in Holland and always longed to get back to the land, while Erik was fascinated with rum because of the Dutch history with sugar plantations in the Caribbean.

In 1999, they bought a farm outside Richland. They played around with a few things, then decided to focus on sugarcane so they could make rhum agricole—rum from fresh sugarcane—as a hobby. Sugarcane actually can grow as far north as Georgia. It does well in the hot, humid summers as long as you protect the roots during winter freezes.

"It was basically two dreams coming together," she says.

They didn't intend for their rum to be a business, but they applied for a license just to be legal. That caught the attention of the mayor of Richland, who asked them if they'd consider opening a distillery in his town.

"We said, 'no—there's nothing there,'" Karin recalls. "And he said, 'That's why—there's nothing there. I want you to help revive the town.'"

So in 2011, they gave it a try, first taking over a columned building that had once been a jewelry store and turning it into a tasting room. Then they bought the pharmacy and the hardware store on either side, turning them into barreling rooms.

Today, they have a string of five buildings, once abandoned, that house their barreling rooms, their two copper pot stills, and their fermentation room. An open space between the buildings has been turned into a pleasant courtyard for visitors to sit and enjoy.

Their rums, made from cane syrup they boil down from their own sugarcane (they also sell the syrup bottled as "Almost Rum") are winning accolades, including rum competitions in the Caribbean. The afternoon I stopped in, Karin had to stop our conversation for a few minutes to take a phone call from an Atlanta businessman who wanted to export their rum to Jamaica.

"As we say in Dutch, that's like carrying sand to the beach," she said, laughing. "But OK!"

Between growing, cutting, and boiling sugarcane and then the work of fermenting cane syrup for five or six days and distilling it into rum, their distiller, Roger Zimmerman, likes to say he only works half-days—from 5 A.M. to 5 P.M. He's joking, of course: sometimes, it's even longer. He has a house nearby, so he's often at the distillery late at night to check on the stills. They were also in the process of opening a second distillery, in tourist-heavy Brunswick, Georgia, where they'll make only white rums.

Back in Richland, their gamble is starting to pay off for the town: there's a microbrewery now, Omaha Brewing, and a farm nearby growing olives and making olive oil. A chocolatier was getting ready to move down from Atlanta, and Karin Vonk has started an April food festival, Taste of Richland, to feature the food from local farms.

The only problem for visitors, she says, is a lack of places to eat. But she's working on it. When I was there, the Vonks had just bought an old hotel and were trying to find a chef to open a restaurant in it.

"That's my next hope, dream, plan," she says.

Even in rural Georgia, the Vonks haven't had any problems with people objecting to what they do, she says.

"The locals are ecstatic that somebody had faith in the town."

304 N. Dudley St., Americus, Ga.; 229-924-3310, thirteenthcolony.com.
Tours offered quarterly; contact the distillery for more information.

Considered by boozehounds to be one of the most significant distilleries in the state, Thirteenth Colony isn't an easy place to tour. Operating with a very small staff on a side street off Americus's downtown, it only offers tours four times a year. To find out when a tour is coming up, you either have to be persistent in emailing or watch the Facebook page closely.

On the other hand, if you're visiting Richland Rum, Americus is only thirty minutes east, and it also gives you a chance to stay in the Windsor Hotel, a magnificent Victorian-era landmark that's famous all over the South. If you can't get into Thirteenth, you can find almost everything they make in the bar at the Windsor for $5 to $6 a shot when I was there—love those small-town prices.

There's another reason to make the effort: Thirteenth Colony (variously spelled 13th) is making one of the best small-batch bourbons in the South. While I was debating whether to include a distillery with such limited tours in a book on touring, I picked up one of their distinct squatty-shaped bottles at a liquor store in Atlanta. One taste and my mind was made up: no list of Georgia distilleries would be complete without them.

After repeated requests, distiller Graham Arthur finally agreed to meet me at the small industrial warehouse on a cold Saturday morning.

"Pretty humble little place," he said, unlocking the door. Humble and little from the outside, true, but on the inside, it's like Platform Nine and Three-Quarters in the Harry Potter books: once you walk down a little hallway and through the tiny front room, there's a vast space behind it that you'd never guess exists from the outside. And it's loaded with racks of full-size oak barrels.

Owner Alton Darby was in real estate when he got interested in the idea of making whiskey. Graham Arthur, in his late thirties, wasn't a distiller. He was a chef in Atlanta, and he was struggling in the 2008 economic downturn.

"It was a horrible time to be a chef," he says. "Nobody wanted to spend money on a plate of my food."

When he heard what Darby was planning, he told him he was ready to leave Atlanta to join him: "'Let me quit my three jobs and I'll come sweep the floors.'"

Learning to make whiskey was "a learn-as-you-go type thing. Trial, error, just a lot of playing."

Their custom-made still is unusual. It's stainless steel on the outside, copper-lined on the inside, with copper rods they can insert when they're cooking whiskey.

Arthur likes to call their process "making grits."

"It's the exact same process, but everybody is doing it differently. You're boiling corn, you're fermenting it, you're boiling the alcohol out."

They tried a rum, but Arthur admits, "it was awful." So they're sticking to vodka, gin, and several kinds of whiskey, including sour mash, rye, corn, and bourbon. The bourbon is aged for four years, everything else for at least two.

They're also doing experiments, what Arthur calls "single-barrelly stuff." When I was there, I made a stop at a local liquor store to grab a bourbon that had been finished in apple-brandy barrels. It didn't taste like apples, but it had an elegant smoothness.

Arthur doesn't regret leaving his apron and the hectic Atlanta restaurant scene behind for a completely different life in Americus.

"I like a nice, quiet, liquor-making lifestyle."

7 FISH HAWK SPIRITS

16162 SW 44th St., Ocala, Fla.; 352–445–1292, www.fishhawkspirits.com.
Tours by appointment. Tasting rooms also at 1600 E. 8th Ave., Tampa,
813–930–5133; and 21 SW 2nd St., Gainesville, 352–792–6699.

You can find Fish Hawk's creations at tasting bars in Gainesville and Tampa's Ybor City. But if you want to visit the distillery, you have to drive way out in the country, down rural highways lined with live oaks draped with Spanish moss and past fields of cattle grazing with the ever-present egrets at their feet. Then you turn into a small neighborhood and keep going, until the paving gives out and you hit a dirt road. Then you keep going some more.

Co-owner and distiller Mike Bogdenovich never expected people to make the trip, until people started turning up. When I was there, he was busily building a tasting room for them.

"When we started, it wasn't logical to have retail," he says. "We didn't anticipate any visitors at all."

There is a reason the distillery is located in such an out-of-the-way spot:

If you make your way out in the country near Ocala, Fla., you can find Fish Hawk owner Mike Bogdenovich smoking oats in an old barbecue smoker.

it's two miles from Rainbow Springs, a massive source of exceptionally pure water Bogdenovich calls "classic whiskey-making water."

The road that led Bogdenovich, fifty-six, here is a long one, too: he was in international insurance, living all over the world. In 1995, he was in Mexico City and started making wine with tropical fruit as a hobby. A lot of his friends were expatriates, including British ones who had worked at distilleries. He got intrigued and bought a fifteen-gallon still, playing with things like carambola.

"Alice down the rabbit hole," he calls it, getting more and more interested in what you could turn into alcohol.

A few years later, his son was getting older and needed to be in America for schooling. The family resettled in South Florida and Mike started searching for what to do next. The brewpub scene was just getting started when he read an article in the *Wall Street Journal* predicting that distilling was going to take off next.

"The penny dropped," he said. He realized, "There's going to be a wave of this."

South Florida has terrible water, but he had an uncle near Ocala, so he found a site there with great water and zoning for small farms. Teaming up with partners, including co-owner David Molyneaux, they got the sixth distilling license in the state.

Fish Hawk's specialty is using as much of Florida's natural produce as possible, like their Francis Marion brandy made from tangerines and a dozen vodkas flavored with Florida fruits, like blueberries. His spicy rum really is spicy, flavored with serrano peppers.

"One rule is, we won't use artificial color or flavor," he says. "The other is, we do business with our neighbors. If we can't find it in Marion County, we try to find it in Florida. If we can't find it in Florida, we try to find it in the Southeast. If we can't find it in the Southeast, we try to find it in America. If it's not available in the U.S., we don't use it."

Among all the unusual things being made at Fish Hawk, two really stand out. One is an oat-based whiskey, similar to an Irish poitín.

"It's the grain from hell," he says. The kernels are much harder than dried corn, so he struggled to figure it out, even trying a meat grinder, before he finally came up with smoking it to soften it. He uses a cabinet-style smoker, like the ones some barbecuers use, to smoke 300 pounds of oats at a time over oak and cherry, then mixes it with raw grains and barley he malts on site.

His other eye-catching concoction is an absinthe, colored red with Florida hyacinth.

"I'm too red-necky to make anything green," he jokes. Since the tasting room wasn't finished, I did my tasting in a shed where the ceiling was lined with 20 buckets of Florida botanicals, labeled on the bottom: wormwood, hyacinth, fennel.

In ten years, Bogdenovich expects to be a mid-size distillery. The potential is there, he says.

"It sounds snotty," he says. "Americans, in my opinion, have finally learned to eat and drink. We're finally appreciating things that are crafted. We want to showcase the agricultural products of Florida, and we want it to be a delicious product that's unique."

800 Thirty-First St. S., St. Petersburg, Fla.; 727-914-0931,
www.stpetersburgdistillery.com. Call for tour appointments.

Tucked back in the Thirty-First Street neighborhood, it can be hard to spot the distillery's sign: the building is across the street from a high school, so they keep a low profile. But the brand, Old St. Pete, isn't low-profile at all. Their bottles are showing up all over the state, particularly in the heavy-tourist areas around Orlando.

"We're not trying to be identical, we're trying to be Florida," says Frank Dibling, who runs the operation. "Florida agriculture–ingredient products."

Their products have a variety of names, including Banyan Reserve vodka, Tippler's orange liqueur, Oak & Palm rums, and Old St. Pete gin, whiskey, vodka, and rum; and a wide variety of styles, from coconut rum and spiced rum from Florida sugarcane molasses to a distilled gin flavored with lemon, grapefruit, and orange. They were buying seven-year-old whiskey from the Midwest and blending it with their own unaged corn whiskey, although the plan is to keep adding more of their own whiskey until it's all made there.

Dibling's background was in manufacturing for the auto industry before he moved to Florida to work for Florida Distilling, which makes Cruzan rum. At sixty-two, he was looking for one last challenge before retirement when he was contacted by owner Dominic Iafrate, who had made his money in recycling and landfills. Iafrate wanted to create a business that his grandchildren could grow up to run, and he gave Dibling and his team a lot of room to create what they think will get a local following.

"We're old Florida, small-batch and artisan," Dibling says. "All we claim is that our products are Florida products. We don't claim to be organic, just all-natural."

St. Pete is another distillery that was quickly adding visitor facilities when I was there, expanding to include a tasting room, rickhouse, and parking for tour buses.

"The farm-to-table thing is a big deal," Dibling says. "Alcohol isn't a health drink. But people are into this farm-to-table. They feel more confident because they know where it's made."

A group of local families raised the money to turn St. Augustine's original electric plant and ice plant into a distillery and restaurant.

9   ST. AUGUSTINE DISTILLERY

112 Riberia St., St. Augustine, Fla.; 904–825–4962, staugustinedistillery.com.
Free tours 10:30 A.M.–5:00 P.M. Monday–Sunday.

St. Augustine is a town with one industry: Florida history and the tourists who come to see it. The oldest jail, the oldest schoolhouse, the Fountain of Youth. Every block in town keeps the historical-marker business thriving.

St. Augustine Distillery is making history itself. Less than a decade ago, it would have been illegal to have a tour and a tasting at a distillery at all. When Florida rules on distilleries loosened up, two dozen local families joined in a project to raise the money to save the old electrical plant and start a distillery in it, so it's completely locally owned.

The distillery exists because of ice. Ice was a miracle in nineteenth-century Florida, the only reason the state's massive agriculture industry could exist before commercial refrigeration. The electric plant, built in 1907, supplied more power than the town needed, so an icehouse was added next door.

Today, the distillery is in the electric-plant side, with a tour that plays up the early industrial connection: when you make a reservation at the door, you get a time card that you stamp with an old time clock.

"You're now an employee," you'll be told. "We pay you in booze."

The tour starts in a small museum that's worth a visit: displays go over the Florida agriculture ties (heirloom sugarcane grown organically at a local farm, non-GMO corn, and red winter wheat), the first distillery in the state (Three Chimneys in Ormond Beach, which opened in 1768 on the grounds of a British sugar plantation), and the ties to developer and railroad tycoon Henry Flagler, who had co-owned a distillery in Ohio before he fell in love with Florida. And ice, of course: there's a good display explaining how ice was made.

The tour takes you through the manufacturing area, with a stripping still named Bessie, for the aunt of an employee who was a good cook, and a refining still named Ella, for Ella Fitzgerald (for her refined style of singing).

It ends at a tasting bar where they do a flashy demonstration of a couple of cocktails—the Florida Mule, made with sugarcane vodka, and a gin and tonic made with their botanical gin and their own tonic syrup. Then you land in a big gift shop loaded with glassware and cocktail tchotchkes.

You're not finished, though: the other side of the building is a restaurant called the Ice Plant, a cavernous space with a menu of Florida-themed dishes and two very busy cocktail bars. On Saturdays, the tours fill up very quickly, so get your time card as soon as you arrive and then head up to claim a spot at the cocktail bar while you wait for your slot.

If you roll your eyes (as I did) at the idea of handmade ice, the drinks at the Ice Plant will set you straight. Their piña colada with hand-shaved ice is a revelation, like a fluffy snow cone.

Other Craft Distilleries in Georgia, Alabama, and Florida

GEORGIA

Dalton Distillery. 109 E. Morris St., Dalton; 706–483–2790, www.dalton distillery.com. Tours 10 A.M.–5 P.M. Monday–Saturday.

Dawsonville Moonshine Distillery. 415 Hwy. 53 East, Dawsonville; 770–401–1211 or 706–344–1210, www.dawsonvillemoonshinedistillery.com. Tours 10 A.M.–5 P.M. Monday–Saturday, 12:30–5:00 P.M. Sunday.

Granddaddy Mimms. 161 Pappy's Plaza, Blairsville; 706–781–1829, www.mimms

moonshine.com. Moonshine distillery and museum. 11 A.M.–6 P.M.
Monday–Saturday, 1–6 P.M. Sunday.

Independent Distilling Company. 731 E. College Ave., Decatur; 678-576-3804,
www.independentdistilling.com. A small distillery making Hellbender
bourbon and corn whiskey, Independent white rum, and Outlier American
single malt in the historic neighborhood near downtown Atlanta. Tour
times usually noon–5 P.M. Saturday.

Stillhouse Creek Craft Distillery. 2148 Town Creek Church Rd, Dahlonega;
706-348-2019, stillhousecreekdistillery.com. Free tours and tastings
11 A.M.–5 P.M. Wednesday–Saturday, 12:30–5:00 P.M. Sunday.

Still Pond Distillers. 1575 Still Pond Rd., Arlington; 800-475-1193, stillpond
distillers.com. Farm-based winery and distillery in a dry county (so
no samples, although you can sample the wines). Distillery tours are
10 A.M.–5 P.M. Monday–Saturday.

ALABAMA

Avondale Spirits. 4100 3rd Ave. S., Birmingham; 205-203-4546,
www.avondalebrewing.com. In the same building with the Wooden
Goat restaurant on the grounds of Avondale Brewing Company.

Irons Distillery. 2211 Seminole Dr., Studio 2061, Huntsville; 256-536-0100,
www.ironsone.com. Tours and tastings noon–6 P.M. Wednesday–Thursday,
noon–8 P.M. Friday, noon–6 P.M. Saturday.

Wolf Creek Distillery. 25020 U.S. 98, Elberta; 251-943-4853, www.wolfcreek
distilleries.com. Launched in 2016, making whiskey and sweet potato
vodka; tours available by appointment.

FLORIDA

Black Coral Rum. 1231 W. Thirteenth St., Riviera Beach; 561-766-2493,
www.blackcoralrum.com.

Drum Circle Distilling. 2212 Industrial Blvd., Sarasota; 941-702-8143,
www.drumcircledistilling.com. Free tours of rum distilling; noon–5 P.M.
Tuesday–Saturday. Tour hours vary; contact the distillery.

Florida Cane Distillery. 1820 N. Fifteenth St., Tampa; no phone number
available, cane-vodka.com. Free tours at 6:30 P.M. Friday and 2 P.M.
and 3 P.M. Saturday and Sunday. Microdistillery making flavored vodka,
moonshine, and gin in Tampa's Ybor City.

Key West Distilling. 524 Southard St., Key West; 305-295-3400, kwdistilling
.com. Free tours on a walk-in basis as long as staff is available; 10 A.M.–5 P.M.
Monday–Saturday. Appointments requested for groups larger than six.

Palm Ridge Reserve. Umatilla; 352-455-7232, palmridgereserve.com.
Microdistillery on a cattle farm, making small quantities of bourbon-style
whiskey. Open for tours a couple of times a month. Contact the distillery
for information.

Peaden Brothers Distillery. 382 N. Main St., Crestview; 850-306-1344,
peadenbrothersdistillery.com. Tours noon–5 P.M. Wednesday–Saturday.
Family-owned distillery making moonshine and flavored corn whiskeys in
a restored 1945 theater in the historic downtown.

Rollins Distillery. 5680 Gulf Breeze Pkwy., Gulf Breeze; 850-503-1275.
Noon–5 P.M. Monday–Saturday. Making vodkas and rums. Call the
distillery to book tours.

South Florida Distillers. 2612 S. Federal Hwy., Fort Lauderdale; 954-541-2868,
www.southfloridadistillers.com. Making Fwaygo rums. Noon–5 P.M.
weekdays; Saturday and Sunday by appointment.

Wicked Dolphin Artisan Rum. 131 SW Third Pl., Cape Coral; 239-242-5244,
wickeddolphinrum.com. Free tours available on Tuesdays, Thursdays, and
Saturdays; reserve online.

Winter Park Distilling Company. 1288 N. Orange Ave., Winter Park;
wpdistilling.com. Brewstillery tours are $11; contact the distillery to
arrange for a tour.

SOUTHERN MULE

From ASW Distillery in Atlanta. A real craze has erupted for Moscow Mules and their signature copper mugs, designed to keep the drink very cold. The loose recipe lends itself to a number of craft spirits.

MAKES 1 DRINK

4 ounces ginger beer
1 ½ ounces whiskey
2 dashes bitters
¼ of a lime

Fill a rocks or old-fashioned glass (or a copper mug, if you happen to have one) with ice. Add the ginger beer, whiskey, and bitters. Squeeze the lime wedge into it and stir gently to mix. Serve very cold.

THE PORCH SWING

From John Emerald Distilling Company, Opelika, Alabama.

MAKES 1 DRINK

1 ½ ounces spiced rum
About 3 ounces lemonade
Basil leaf and basil sprig

Fill a rocks glass with ice. Add the rum, then top with lemonade. Bruise a basil leaf (just crush it lightly in your fingers) and add to the drink. Stir and garnish with a basil sprig.

DEEP SOUTH COSMO

From Lazy Guy Distillery, Kennesaw, Georgia. They suggest their unaged corn whiskey, Cold Heart, but any unaged corn whiskey will work.

MAKES 1 DRINK

1 $\frac{1}{2}$ ounces unaged corn whiskey
$\frac{1}{2}$ ounce triple sec
1 ounce cranberry juice
Orange slice and maraschino cherry

Fill a shaker with ice and add the whiskey, triple sec, and cranberry juice. Shake well and strain into a chilled cocktail glass. Garnish with an orange slice and a cherry.

MGP: Right or Wrong?

A few years after the cult of bourbon took off around 2010, a controversy sprang up: some of the brands most sought after by enthusiasts, such as Angel's Envy, Bulleit Rye, High West, Smooth Ambler, and Redemption, weren't being made by the distilleries selling them. They were being made at a large facility, MGP, for Midwest Grain Products, in Lawrenceburg, Indiana.

MGP is an old company that started in 1847 and changed hands several times over the years, from Seagram in the 1930s to Pernod Ricard, which sold it in 2007 to a holding company. When that company collapsed in 2011, it was bought by MGP Products, a Midwestern company. The facility in Indiana doesn't release its own brands, instead making whiskeys in a variety of styles but mostly rye, with a huge number of barrels. Companies can buy whiskey they bottle under their own labels or pick the flavor characteristics they're aiming for in what they'll eventually produce themselves.

As the word got out in the whiskey world, it was written about in scathing articles, either expressing outrage or cynicism about the growth of marketing over facts. (Writer Charles Cowdry dubs distillers that buy from MGP "Potemkin distilleries" because they make it look like they're something they're not.)

In some cases, the charges were well-founded: the wording on many bottles didn't make the origin clear, leading to suspicion about how a lot of American whiskey was being made. In a class action settlement, Templeton Rye was forced to change its label to make it clear it was made in Indiana and wasn't "small batch" from "an old family recipe" at all.

There is another side to it, though, that you'll encounter in the craft world. In order to survive their start-up years, small distilleries have to have something to sell, and vodka and corn whiskey often won't cover the costs of buying equipment and getting licenses. Some work with other distilleries, including MGP, to choose whiskey they can sell while they wait for their own barrels to ripen. Some buy whiskey and then re-barrel it themselves for shorter times than the two to four years they have to wait for their own on-site creations.

While the idea of a large distilling facility smacks of factory production, MGP has a respected master distiller and makes very good whiskey. Those sought-after brands being made there got popular for a reason. The issue wasn't quality, it was labeling, marketing, and the consumer's right to know.

Some distilleries, notably Smooth Ambler, have now adopted transparency policies using phrases like "whiskey curation." New Riff Distilling in Kentucky, located just across the bridge from Cincinnati, calls its MGP-made bourbon "O.K.I.," for "Ohio Kentucky Indiana," while ASW Distillery in Atlanta calls the whiskey it gets from a Charleston maker and re-barrels in MGP barrels "Fiddler's," because "we buy it and fiddle with it."

There's a similar but separate issue with distilleries that buy neutral grain spirits from ethanol makers in other states and flavor it themselves for their vodkas and gins. That may be a bigger problem, since they don't control the quality of the base liquor they're using. It's a good thing to ask about when you tour a distillery. I didn't encounter any that weren't up front about it when I asked.

Ultimately, it's up to the consumer to decide if these practices are OK or not. If you're concerned about it, ask—and look carefully for "made in Indiana" on labels.

6

Louisiana and Mississippi

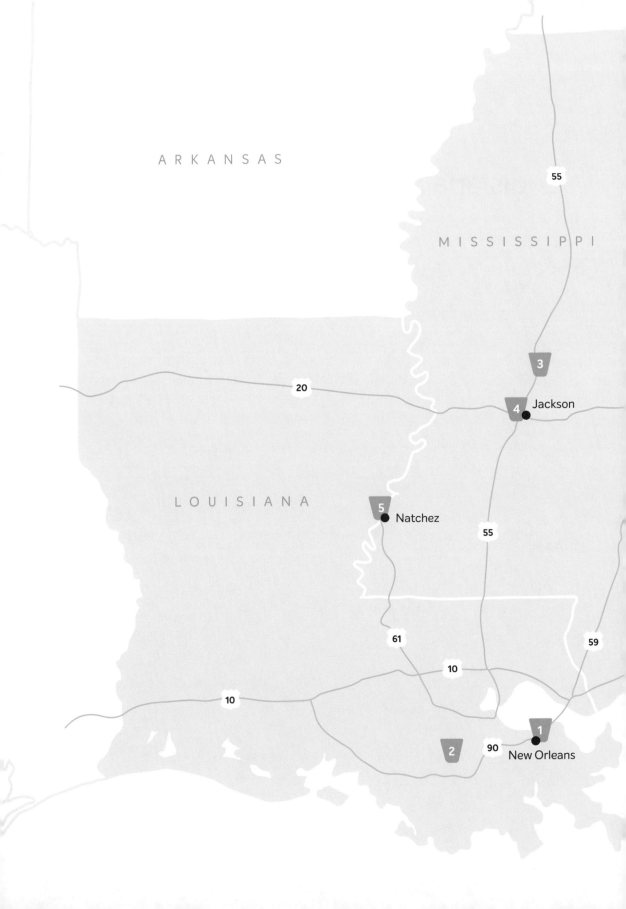

ARKANSAS

MISSISSIPPI

55

3

4 ● Jackson

20

LOUISIANA

5
● Natchez

55

61

10

59

10

2

90

1
● New Orleans

Louisiana and Mississippi

1   Old New Orleans Rum, New Orleans, La.
2   Donner–Peltier Distillers, Thibodaux, La.
3   Rich Grain Distilling Co., Canton, Miss.
4   Cathead Distillery, Jackson, Miss.
5   Charboneau Distillery, Natchez, Miss.

ALABAMA

IT'S NO SURPRISE, for Mississippi, anyway, that this trail is the shortest in the book. Mississippi's restrictive rules on alcohol are legendary: Prohibition started statewide in 1908, a full twelve years before it took effect nationally in 1920, and wasn't repealed until 1966. Of the state's eighty-two counties, twenty-five are still dry.

With those numbers, and with laws that still prohibit distilleries from selling bottles or cocktails on site, it's a wonder that even three distilleries had managed to navigate their way into business by 2016.

The small number of distilleries in its neighbor, Louisiana, is more surprising. The prevailing attitude of "laissez les bon temps rouler" brings the party-minded from all over the world. In addition, the miles of sugarcane fields that blanket the state make sugar one of Louisiana's primary products. And where's there's sugar, there usually is rum.

Besides, New Orleans has one of the nation's most vibrant food cultures, and that's not a recent development—it's been known as the place to dine since the early days of American history, and its Peychaud's bitters played a role in cocktail history. So wouldn't Louisiana be the natural spot for a hotbed of American alcohol-making?

"It's a strange thing," says Gavin Cress, the distiller at Old New Orleans Rum Distillery, officially the oldest rum distillery in the United States—it marked its twentieth anniversary in 2015. "You'd think it would be many more, but it's almost nonexistent."

One reason, Cress thinks, is the closeness of the Port of New Orleans to the Caribbean and Central America: it's cheaper to make liquor down there and easy to get it into the United States, so the state's massive sugar industry didn't have any motivation to make it into rum.

Still, craft-rum distilling in Louisiana is starting, following a historic path. After all, if it weren't for Christopher Columbus, there wouldn't be sugar in the New World at all. Grown for thousands of years in Southeast Asia, sugarcane had made it as far as the Canary Islands. That's where Columbus found it on his second voyage, in 1493, and brought sugarcane shoots and slaves who knew how to grow it to the Caribbean.

Sugarcane loved the hot, humid islands, and it became a major industry, dominated by the Dutch and the British. Check the routes of slave traffic from West Africa and it's astonishing to see the sheer number of ships that brought slaves into the West Indies and the Caribbean for the sugar industry, far more than the ships that brought human cargo to the American colonies to work in cotton and tobacco.

It took another 200 years before people figured out that molasses, the dark, smoky goo left when you boil sugarcane juice to make crystals of sugar, had another use besides baking with it or dumping it: if you mixed it with water and yeast, you could ferment it and distill it.

By the 1700s, there were hundreds of rum distilleries in the United States. Most still weren't in the South, though. Molasses for rum was mostly shipped to Boston for sale to distilleries in New York, Massachusetts, and Rhode Island, which made rum one of the most popular liquors in America. While some reports claim it lost popularity because of anti-British sentiment during the Revolutionary War, that may be exaggerated. George Washington included Barbados rum during the celebration of his inauguration in 1789. It's more likely that the rise in popularity of corn- and rye-based whiskeys simply supplanted it.

Still, rum came to America long before sugarcane did. It was more than 300 years after Columbus brought sugarcane to the Caribbean until it finally became a crop in Louisiana. In the 1790s, a blight coupled with several bad harvests wiped out the indigo crop that had been a major money-maker for the state's plantations, leading planters to replace it with sugarcane and eventually making sugar one of the state's primary industries.

Today, craft rum is finally gaining a small toehold in both Louisiana and Mississippi, along with bourbon and vodka. In the spring of 2017, as many as five new distilleries had either opened or were under construction in New Orleans, in addition to the long-established Old New Orleans Rum. And Louisiana liquor tourists do have an easier time than in Mississippi. While state law won't yet allow sale of cocktails at distilleries, you can buy twelve bottles per spirit per month. Let the good times roll, indeed.

One interesting thing about touring distilleries in Louisiana and Mississippi: with a few exceptions, the landscape around you isn't as dramatic or exciting as the mountainous areas of the South. But the people certainly are. From the bayou to the banks of the Mississippi River, visiting distilleries will bring you into contact with people whose love of legend, history, and tradition will make it a memorable trip.

For this trail, you could start at the top of Mississippi and work your way south. Or you can do what I did: fly into New Orleans and start there, make a quick detour southwest into the bayou, then drive north to the edge of the Mississippi Delta and work your way south again, ending up back at Louis Armstrong International Airport.

1   OLD NEW ORLEANS RUM

2815 Frenchmen St., New Orleans; 504-945-9400, oldneworleansrum.com. Tours are $15, including a complimentary cocktail. Free shuttle bus pickup at the French Market (behind Organic Banana) 11:30 A.M., 1:30 P.M., and 3:30 P.M. Monday–Saturday, 1:30 and 3:30 P.M. Sundays, or you can drive to the distillery on your own.

Founded in 1995, this is the oldest distillery in the book and the only one that exceeds the fifteen-year standard I used to define craft distilleries. But its limited production, small staff, and the role it has played in Louisiana still puts it firmly into the mold of a craft distillery.

The distillery is in an industrial area in the Chantilly neighborhood, about fifteen minutes from the French Quarter. If you don't want to drive it yourself, sign up for the shuttle that picks you up in the back of the French Market, in the open-air end several blocks down from Café du Monde. You'll get picked up in a beat-up tour bus with "Gingeroo" and "Rum for Your Life" spelled out in duct tape. (It's New Orleans—just roll with it.)

Once you get to the distillery, it gets better. Inside, you'll find a comfortable tasting room and they'll welcome you with a free cocktail. (On the day I was there, it was a choice of one they called Better Than a Hurricane and another based on peach/mango tea.) The room is decorated with colorful paintings of New Orleans houses and street scenes, painted by founder James Michalopoulos.

Michalopoulos didn't set out to be a rum distiller. He was, and is, an artist. On a trip to Switzerland in the 1990s, he was at a party when the host started pouring his own homemade liqueur. Michalopoulos was bitten by the bug of making his own booze.

Back home, he started asking his artist friends for advice on what he should make. They all agreed: New Orleans was surrounded by sugarcane fields, but no one was making rum out of it. It actually was still illegal when he started, beginning with silver rum, and then putting some of it into used bourbon and cognac barrels to age. When the distillery celebrated its twentieth anniversary in 2015, it created a premium brand, King Creole, combining batches that are eight, thirteen, and seventeen years old.

The distillery started out as Celebration (that's still one of the signs on the building) and changed names a half dozen times before it finally became Old New Orleans.

In the distillery behind the tasting room, where the floods from Hurricane Katrina reached eight feet and wiped out half their barrels, you'll see an unusual still. Instead of a copper pot, they repurposed old dairy tanks, with copper tubing leading to a cooling contraption built from two stacked fifty-three-gallon barrels. It gets the job done, my tour guide assured us. And it was a lot cheaper. It's New Orleans, after all: just roll with it.

2 DONNER–PELTIER DISTILLERS

1635 St. Patrick St., Thibodaux, La.; 985-446-0002, www.dpdspirits.com. Free tours at 4 P.M. Monday–Friday, 2–4 P.M. Saturday (or by appointment for groups of eight or more); tasting room 10 A.M.–6 P.M. Monday–Friday, noon–6 P.M. (prices vary).

Drive an hour outside New Orleans and you're deep in the bayou, where green algae-covered swamps lurk just a few feet from the road in places. (Don't stray far from your car—alligators can move fast on land.)

When you reach the small farm town of Thibodaux, the cane fields stretch off for miles. Thibodaux is still so connected to the French-speaking Acadian culture that the town has French food festivals every year and several B&Bs have French-speaking owners. At Donner-Peltier, the maker of Rougaroux rums and Oryza vodka and gin, they get so many French visitors that they keep a script of the tour translated into French so non-English speakers can follow along. (Oryza is the Latin word for rice, by the way. And Rougaroux … well, I'll get to that.)

Taryn Clement, the marketing manager, is in her twenties and a native of Thibodaux. Her sentences are sprinkled with the distinctive local accent, like *pee-CAWN* and *prah-LEEN*.

"It comes out stronger when I'm drinking," she jokes. "Then you need subtitles."

The founders of the distillery are both doctors, neurosurgeon Tom Donner and pediatrician Henry Peltier. Both grew up in Thibodaux and came back to

If you drink too many samples at Donner-Peltier Distillers in Thibodaux, La., you actually can turn yourself into the legendary Rougaroux—for a few minutes, not the 101 nights of the actual curse.

practice medicine. On a trip to Coeur d'Alene, Idaho, they toured a rum distillery and started to wonder: their town was surrounded by sugarcane, and Peltier's family had a share in a sugar mill. Why was no one making rum?

"There's literally sugarcane across the street," says Clement.

Working with three stills—Betty, Veronica, and Stella—they make their neutral grain spirit for their vodka from corn and long-grain rice grown in Louisiana. Their 96-proof gin is favored with juniper, lavender, pink peppercorns, coriander, orris, angelica roots, grains of paradise, cantaloupe, orange and lemon peel, and Satsuma oranges from a tree in her grandfather's backyard.

They're also barrel-aging rums, although they can only keep them for two years. The extreme heat and humidity in summer and the wide temperature swings in winter, where it can go from twenty degrees to eighty and back to twenty in a week, hurries aging so much that the rums get too woody if they leave them long.

Now, about that name "Rougaroux." Part of the lesson at Donner-Peltier isn't just how to make liquor. You also have to learn about the Rougaroux.

The French word for a werewolf is loup-garoux. In the bayou, it got twisted a little, and it got a new life.

"Around here, everybody either works in sugar or grew sugarcane," says Clement. Her own family had a nine-acre field right behind the house.

Children love to explore, but a sugarcane field is a dangerous playground. There are snakes and wandering alligators, the leaves are sharp as razors, and the tall stalks make it easy to get lost. To keep children out of the fields, parents would tell them the story of the Rougaroux.

A human cursed by a Rougaroux turns into a fierce creature under the full moon, with a human body, the head of a wolf, and six fingers on each hand.

"If you see one, you're about to get cursed or eaten," says Clement. In the daylight, if you see a person with six fingers, that's a cursed human: for 101 days, under every full moon, that person will become a Rougaroux unless they can pass the curse on to another person.

Donner-Peltier's rum labels feature a stylized stand of sugarcane stalks with two eyes peering out, and the names are all references to the legend: Full Moon is dark and aged with oak chips for more flavor, Sugarshine is a 101-proof clear rum. Their praline rum isn't as sweet as you'd expect—they steep it with local pecans, so it has a creamy/nutty flavor—and the name, 13 Pennies, is another part of the legend. If you look outside the door of the distillery, you may spot thirteen pennies lined up.

To help frightened children sleep on full-moon nights, parents line up thirteen pennies outside. If a Rougaroux sees the pennies shining in the moonlight, it will stop to count them. But because it has twelve fingers, it gets confused when it reaches thirteen and starts over, keeping it occupied and out of your house until the sun rises. Sort of like someone on Bourbon Street after a few daiquiris made from 101-proof rum.

TRAVELER'S NOTE: This is a great part of the world to eat, but don't look for fancy. Instead, drive to the other side of town from the distillery and find the Bourgeois Meat Market ("Miracles in Meat Since 1891"). They specialize in excellent smoked beef jerky and a spicy version of head cheese. There's also a hot box where you can usually find juicy boudin burritos, the perfect thing to eat while you drive back through the bayou.

339 W. Peace St., Canton, Miss.; 601–391–3190, www.richdistilling.com.
Free tours on Saturday, by appointment. Contact the distillery.

It's a three-and-a-half-hour drive up to northern Mississippi, with a chance to get off and drive part of the way on the historic Natchez Trace, the old road between Natchez and Nashville.

Canton is a very small town, but it was once an active place for liquor. As a railroad town, hotels and saloons around the town square used to buy plenty of whiskey from big distilleries and bottle it as their own brands.

David Rich, thirty when I met him, is a genial young man who used to be an engineer. At the University of Mississippi, he got seriously interested in bourbon. After he graduated and got a job in Huntsville, he just couldn't shake his bourbon passion.

"I liked my job," he says. "I made a lot more money. But I wanted to work for myself and I wanted to move home."

Once Cathead Vodka in Jackson figured out how to navigate the state's liquor laws, he figured it could be done. So he came home to Canton, bought an 1880s building that used to be a cotton gin and gristmill, and set out to make bourbon the way he thought it should be made.

For the name, he was making lists of every word he could think of that was related to the South and whiskey, from "moonlight" to "magnolias." One night, he was searching a site on whiskey-making history and decided to plug in his own name, Rich. He turned up the logo and name for Rich Grain, a Kansas City distillery that disappeared during Prohibition. The name wasn't trademarked, so he grabbed it.

Mississippi's laws have limited his growth. He's got a nice tasting room with a copper-topped bar and heart-pine floors all ready to go if he ever gets to sell cocktails and bottles of his corn whiskey, bourbon, and rum. So far, all he can do is offer quarter-ounce tastes. To buy bottles, you have to go to Jackson or other nearby cities with liquor stores.

Still, he's finding plenty of support in town. Canton has a big Victorian Christmas festival every year, as well as flea markets and antique shows. They all bring in visitors.

Despite the state's restrictive rules, he hasn't run into any resistance in the town.

"Around here, it's been incredibly positive. They understand this isn't a den of iniquity with women dancing on the bar."

David Rich of Rich Grain has set up his small distillery in Canton, Miss., so that it's an efficient operation for one man. An engineer by training, he built his still himself using copper tubes from an old TV station's antenna.

As a one-man distillery, Rich's still setup is very efficient, sort of like a galley kitchen with everything within arm's reach. Since he's an engineer, he wanted to build his own still. The copper pipes and tubes used to be part of the coaxial antenna at a TV station that was being shut down in Seattle, Washington. He even has a tabletop mill for grinding his own corn.

"I do everything from scratch in here," he says. "That's the point of craft, right? You should be doing it all from scratch."

4   CATHEAD DISTILLERY

422 Farish St., Jackson, Miss.; 601–667–3038, catheaddistillery.com.
Tours on the hour from 3–6 P.M. Thursday, 3–7 P.M. Friday, and 1–7 P.M. Saturday.

The first distillery to crack Mississippi's liquor laws, Cathead is ready to party. Originally opened in 2010 in the small town of Gluckstadt, it moved into a large warehouse on the edge of downtown Jackson in 2015.

With a huge single room, there's an open area of tables scattered with board games, a beer bar with craft brews on tap, a small tasting bar for their vodkas, and distilling tanks and a barrel room in back. There are even corn-hole boards all lined up and ready to go, and there's often live music at night.

The story of Cathead starts with music: owners Richard Patrick and Austin Evans went to the University of Alabama together and bonded over a mutual love of blues. When they decided to go into business making liquor, they took the name Cathead from a blues term. It was a phrase that musicians used when they liked the playing of another musician, and it also was sometimes used for folk-art figurines made from mud from the Yazoo River. (It's also gotten popular with food fans as a large, rough form of a drop biscuit.) A portion of Cathead's sales go to programs to support struggling musicians.

With a column still and a pot still, they're making their own neutral grain spirit for lightly sweetened vodka in several flavors, including honeysuckle and pecan, plus gin, and Hoodoo, a peppery hickory liqueur. They're also working on bourbon that's still aging.

When I visited, the tour was led by young people following a pretty cursory script. If you already know the basics of distilling, you won't learn much. But Cathead's facility is a comfortable place to visit, and its role as the start of craft distilling in the state deserves a nod.

They have distribution in fourteen states, which a tour guide simply explains as "The SEC [Southeastern Conference] and Colorado. People in the South get that."

5   CHARBONEAU DISTILLERY

619 Jefferson St., Natchez, Miss.; 601-861-4203, www.charboneaudistillery.com.
Tours hourly 5–8 P.M. Friday and noon–6 P.M. Saturday.

If you want an experience of the Old South, there aren't many places more interesting than Natchez. Located high on the bluffs of the Mississippi River, it's as much Louisiana in spirit as it is Mississippi, with blocks of historic antebellum homes.

Charboneau Distillery is particularly interesting to visit, because it's sort of founded on hospitality. Chef Regina Charboneau is famous as a hostess, with a bed-and-breakfast, Twin Oaks, in an immaculately restored 1830s mansion. Downtown, she runs a cocktail bar and restaurant, the Kings Tavern, in a building that dates to at least 1780.

Next door, in an old dry-goods store, her husband, Doug, and their son Jean-Luc have started a rum distillery that's the culmination of a dream that's just a little older than Jean-Luc himself, who was twenty-seven when I met him.

Regina and Doug started out in California, where Regina had a San Francisco restaurant and Doug, a business consultant who specializes in saving troubled companies, also worked with several wineries. When they were getting ready to start a family, they took a "last hurrah" trip to Martinique. Doug wasn't a big fan of rum until they were served a twenty-one-year-old rum. He was stunned: "My God, where has this been?" They loved it so much, they started collecting aged rums and fantasizing: "Someday, we'll have a rum distillery when the children are grown."

In 2000, they moved to Natchez, Regina's hometown, and started the inn, attracting a following from some of their famous friends in California. Guests have included Anderson Cooper, Allison Janey, and even Mick Jagger.

Doug divided his time between Natchez and New York, but he was ready to settle down. Jean-Luc, the oldest of their three children, was working in New Orleans at restaurants when Doug called him and told him it was time to come back to Natchez and help him open a rum distillery.

When they wanted to buy the corner store for the distillery, the old tavern was a part of the property. So Regina turned it into a restaurant with a great tavern menu (don't miss the crawfish pot pie), and Jean-Luc runs both businesses.

Your tour starts in the tavern with a rum cocktail, and then you can return there to grab lunch or dinner and shop in the bottle and gift shop upstairs; under Mississippi rules, you can't sell cocktails or bottles in a distillery. But since the Charboneaus have two buildings, they can have separate licenses that let them own a cocktail bar, a bottle shop, and a distillery. One of their bartenders calls it "the strangest mini-mall in Mississippi."

In a weird way, Doug's analytical mind has helped with learning to make rum. "Science vs. math: it's a different side of the brain," he says.

When Doug ordered a copper still, it came with a computer display that he didn't think he needed. But he realized it would help, so he called a friend who does programming to come set it up. It turned out that the friend was also a rum geek. They sat down with more than twenty bottles, tasting them to decide what profile Doug liked. Once they picked a style, they went to a sugar mill and worked out a recipe to get that result, using a combination of molasses and raw sugar. After starting with a white rum, he started putting

The Mississippi River was once a major route for the growth of southern whiskeys and bourbons, and it still leads the way through craft distilling.

some in small barrels, lined up under a window, to age just a little while, for gold rum, and others that are aging longer.

When you finish your tour, tasting, and a little shopping at Charboneau, get in your car and head west about three blocks to a small park on the high bluffs above the Mississippi River.

If you've followed this book in geographic order, as I mostly did while researching it, you've reached the end of the line.

Before you end your trip, do me a favor: stand at the edge of the Mississippi River for just a moment. Try to imagine it as it was in the 1800s, busy with flat boats loaded with barrels of whiskey from Kentucky and paddleboats crowded with new Americans coming from all points north into the heart of the South, bringing all their drinking tastes and traditions with them.

It all passed by here, once upon a time.

And now, you've become a part of the journey, too.

Other Craft Distilleries in Louisiana and Mississippi

Atelier Vie. 1001 S. Broad Ave., New Orleans; 504-534-8590, www.ateliervie .com. Making gins, 125-proof vodka, brandy, unaged whiskey, and both red and green absinthes from some ingredients they grow themselves. Free tastings and bottle sales 10 A.M.–2 P.M. Saturdays and Sundays.

Cajun Spirits Distillery. 2532 Poydras St., New Orleans; 504-875-3592, cajunspirits.com. Making Crescent Vodka, Tresillo Rum, and 3rd Ward Gin. Tours not yet available, but they plan to add them in the future.

Louisiana Spirits Distillery. 20909 S. Frontage Rd., Lacassine; 337-588-5800, www.laspirits.net or www.bayourum.com. Makers of Bayou Rum in western Louisiana. Free tours hourly from 10 A.M. to 4 P.M. (except at noon) Tuesday–Saturday. Gift shop and tasting room open from 9:30 A.M.–5:30 P.M.

Lula Restaurant Distillery. 1532 St. Charles Ave., New Orleans; 504-267-7624, www.lulanola.com. The only microdistillery/restaurant in Louisiana, making gin, vodka, and rum on site and serving brunch, lunch, and dinner in the restaurant. Tours are free; tasting flights are $15. 11 A.M.–10 P.M. Sunday–Tuesday, 11 A.M.–11 P.M. Wednesday–Saturday.

NOLA Distilling. 3715 Tchoupitoulas St., New Orleans; 504-518-5545, noladistilling.com. Sweet potato vodka made from locally grown sweet potatoes. Tasting room hours: 3–9 P.M. Tuesday–Thursday, 2–9 P.M. Friday, noon–10 P.M. Saturday, and noon–8 P.M. Sunday.

Rank Wildcat Spirits. 619 Bonin Rd, Lafayette; 337-257-3385, rankwildcat .com. Tours by appointment.

Roulaison Distilling Co. 2727 S. Broad St., New Orleans; 504-517-4786, www.roulaison.com. Making rum from Louisiana sugarcane, with a focus on historic styles. Tasting room hours: 3–6 P.M. Fridays.

MISSISSIPPI

Crittenden Distillery. 19193 Hwy. 43, Kiln; 228-255-2058 (no website, but it does have a page on Facebook). This distillery was still under construction. Check for updates if you're planning a trip in the area.

Recipes

HONEYSUCKLE LEMONADE

Cathead Distillery in Jackson makes an ethereal vodka flavored with the honeysuckle that blooms all over the South in late spring. If you can't get your hands on it, you can also make your own (see note).

MAKES 1 DRINK

2 ounces Cathead honeysuckle vodka
4 ounces lemonade (from scratch or from a mix)

Fill a rocks or old-fashioned glass with ice. Add the vodka and lemonade and stir to combine. Garnish with a sprig of mint or basil.

NOTE: *Alcohol is a solvent, so it makes a good vehicle for picking up the aroma of flowers. Making honeysuckle-flavored white rum or vodka is simple: just get about 2 cups of fresh honeysuckle blooms and place them in a glass bowl. Cover with vodka or white rum and let stand overnight. Strain. The haunting flavor will keep best if you store the alcohol in the freezer.*

PITCHER OF RUM PUNCH

From Regina and Doug Charboneau of Charboneau Distillery and Kings Tavern, Natchez, Mississippi.

MAKES 6 TO 8 SERVINGS, DEPENDING ON SIZE

1 ½ cups gold rum, such as Charboneau
½ cup frozen limeade concentrate
¼ cup frozen orange juice concentrate
4 cups pineapple juice
¼ cup grenadine
3 cups water
Lime slices (garnish)

Combine all the ingredients in a pitcher. Stir well. Serve over ice with a lime twist.

Medals: Does the Bling Mean a Thing?

At many distillery tasting bars, you'll see a display of bottles with more medals hanging around their necks than Olympic swimmer Michael Phelps: bronze medals, silver medals, gold medals, all from a dizzying array of competitions.

Do the medals actually tell you much? That depends on the competition.

Lew Bryson, an experienced writer on spirits and beer, has judged a number of competitions himself. The proliferation of contests can make it hard to know which medals actually carry any weight, and it can be difficult to figure out whether the medal actually tells you anything.

"No one's rating the raters," Bryson says. "When I see double golds coming out, the BS detector starts to ping. It's like 'double secret probation'—what does that mean?"

If everybody wins a medal, the way some contests are set up, then "bronze" can just be the equivalent of those T-ball games where every kid gets a trophy for showing up.

The competitions that make Bryson the most suspicious are the ones that charge high fees to enter, more than just what's required to cover the contest's expenses.

"There's an unhealthy correlation between paying fees and the number of medals handed out," he says. "Unfortunately, with the large number of craft distilleries and really, really small companies with no budget for promotion, it can look like a good spend."

The competitions that mean the most are obviously the ones with the most transparent judging procedures. The annual ratings by Tasting.com, run by the Beverage Tasting Institute, for instance, garner a lot of respect because they use judges who are well-qualified in their specialties, categories are judged one at a time—all the gins at one session, all the vodkas at another—and they use a blind-tasting process with no discussion allowed until all the judges' evaluations are made.

Bryson also likes the competitions run by the two craft associations, the American Distilling Institute and the American Craft Spirits Association, which use judges from within the industry.

"One, I know the people doing the judging and I trust them, and two, there are fees charged but it isn't a money maker, it's an expense-covering thing."

Other really big competitions, such as the San Francisco World Spirits Competition, get a lot of attention, and a win is certainly something to brag about. But Bryson worries about the sheer number of bottles being judged and whether there are enough judges who really know the liquors they're tasting.

"There's a lot of newness in what you're tasting [with craft spirits]," he says. "A wine person does wine really well, but I've seen them tasting whiskey and they don't know their [blank] from a hole in the ground. Some of these younger craft spirits are getting amazing reviews because you don't have a lot of comparison to judge them against.

"There's a whole issue there. Some of this stuff is just completely new. What standards do you judge it against?"

So while the medals are nice and any distiller who receives one will point it out, be careful about giving competitions too much weight unless you know what the competition was and how it was judged.

"When it really comes down to it, you ought to be tasting for yourself," says Bryson. "At least find out who you trust to do the tasting for you."

Alembic. The earliest distilling equipment consisted of two containers, called retorts, connected by a tube. One side was heated, causing steam to rise and travel through the tube to the second container, isolating certain molecules from the liquid being heated. Originally from an Arabic phrase, al anbic, and possibly from a Greek word, ambix. Glass versions are used in laboratory work and metal versions, usually copper, were used in France to make cognac and eventually became the model for the pot still.

Batch distilling. Making one batch of alcohol at a time in a pot still, versus the continuous production of a column still.

Bung. The round wooden core used to close the hole in a barrel. Usually, the bung is made of a softer wood than the oak of the barrel, often birch, so it pops out easily when the barrel is struck.

Cogeners. The substances in the foreshots and heads that are removed before the run crosses over into producing ethanol during the heart of the run. They include methanol, fusel alcohol, acetone, and acetaldehyde, as well as esters and tannins. Not all are poisonous, although many distillers claim those are the things that cause hangovers and they tout the removal of them as keeping their spirits from giving you a headache.

Column still. Also called a continuous still, a patent still, or a Coffey still. A tall tube with perforated plates at set intervals, with a glass port in front of each one. Column stills vary widely in size, from short ones with six to eight ports up to as tall as sixty feet or much higher at large distilleries. Instead of heating the wash, or the liquid from fermentation, directly, as a pot still does, the liquid enters from the side, falling to the bottom where it is heated. The steam rises, passing the plates, where droplets catch and fall back down, then rise again until the droplets of alcohol reach the top. Every time the mixture rises, it's distilled a little more. A column still is more efficient and can run continuously, but they also don't allow the same flavor characteristics as pot stills. Many distillers use a combination of pot and column stills, or two column stills; the second is called a doubler, for double distillation.

Condenser. A second chamber, next to or built into the pot still, where the hot steam of alcohol is cooled, returning to a liquid. You'll see all kinds of styles, from tall copper pots to oak barrels with a copper coil inside.

Foreshots. While many distillers simply refer to this as a part of the heads (see definition), foreshots are the very first alcohol to come from a distilling run. It contains things you can't drink, including acetone, methanols, and ethyl acetates, and must be discarded.

Glencairn. The best glass for tasting whiskey, it has a rounded bottom that rises into a flared top. It helps to trap aromas and deliver them to your nose and palate. Many tasting rooms use miniature versions for tasting.

Heads. The first part of the alcohol separated at the beginning of the run, produced at around 133 degrees. The heads usually contain poisons, including methanol and acetone, and are discarded before the good alcohol is collected.

Hearts. The main part of the run, containing pure alcohol and flavor esters. This is what you collect and either bottle for immediate consumption or put into a barrel to age.

Low wine or low wines. The low-alcohol liquor produced in the first run or stripping run.

Malt. Soaking grain, usually barley, in water until it germinates, and then drying it to retain the enzymes it would use to convert the grain's starch into sugar to feed the young plant. Adding malted barley to a cooked mash of corn, rye, or wheat speeds the production of starch into sugar, which is eaten by yeast to produce alcohol.

Mash. The mixture of grains, usually corn, but also rye, wheat, or barley, that's ground and cooked to release its starch so yeast can eat it and produce fermentable sugar.

Mash bill. The combination of grains used in creating mash. It's the distiller's "recipe," along with the yeast strain and the style of distilling.

Moonshine. It used to mean any liquor that was illegally produced and untaxed. Legal moonshine today usually is unaged corn-based whiskey, although it can be neutral grain spirit.

Neutral grain spirits. Sometimes just called "NGS" in the liquor world, it's the clear liquid you get from distilling a mash. It can be flavored for vodka or gin, or put into a barrel to age into whiskey. Some distilleries buy NGS from ethanol producers in the Midwest and use it as a base to make something else. One of the first things to ask at a distillery (if they don't tell you immediately) is, "Do you make your own neutral grain spirits?"

Parrot. The end of the pipe where liquor emerges from the still. It's often shaped like a little beak.

Pot still. One of the oldest styles of stills. Often made from copper, pot stills can also be made from stainless steel, sometimes with a copper lining. While the shapes, particularly the top or cap, vary widely, it usually has a round bottom, a cap on top and a pipe leading to the condenser. Pot stills are heated from the bottom or around the sides.

Proof. The amount of alcohol in a spirit is measured as either the percentage of alcohol by volume, or by the proof. Proof is twice the percentage of alcohol, so a 100-proof whiskey is 50 percent alcohol by volume; 80 proof is 40 percent alcohol.

Stripping run. The first distilling run, stripping alcohol from the distillate. The alcohol gathered is usually distilled a second or third time, to raise the proof, to make it smoother and to concentrate the flavors.

Tails. The end of a run, after the heads and hearts have been removed. The tails are weaker and can have off-flavors (often described as "stinky feet"). But there are a number of flavors in the tails, and usable alcohol. They may be discarded or collected and added to the next batch to collect more alcohol and more flavors.

Wash. The grain-based mixture that is distilled into alcohol.

White dog. The clear output of a whiskey still before it goes into a barrel for aging.

Worm. The copper coil where the steam of alcohol from a still collects and is chilled, usually with cold water, to return it to a liquid.

RECOMMENDED READING

Distilling and liquor history can be a complex subject. While I did all of my own reporting for this book, I also read extensively on the subject to double-check my own observations and to fact-check what I was told at distilleries. Since so many distillers are self-taught, you'll often hear information that doesn't necessarily match up.

The following are some of my favorite books that I've used for guidance, to shape my thinking, and sometimes to keep me company during long nights in distant motel rooms.

Alcohol: A History, by Rod Phillips (University of North Carolina Press, 2014).

Bourbon Empire: The Past and Future of America's Whiskey, by Reid Mitenbuler (Viking, 2015).

Bourbon: The Rise, Fall, and Rebirth of an American Whiskey, by Fred Minnick (Voyageur Press, 2016).

Finding the Flavors We Lost: From Bread to Bourbon, How Artisans Reclaimed American Food, by Patric Kuh (Ecco, 2016).

Lost Recipes of Prohibition: Notes From a Bootlegger's Manual, by Matthew Rowley (Countryman Press, 2016).

Proof: The Science of Booze, by Adam Rogers (Mariner Books, 2015 reprint edition).

Southern Spirits: 400 Years of Drinking in the South, by Robert F. Moss (Ten Speed Press, 2016).

Tasting Whiskey: An Insider's Guide to the Unique Pleasures of the World's Finest Spirits, by Lew Bryson (Storey Publishing, 2014).

ACKNOWLEDGMENTS

You don't combine fourteen months of researching a travel book with a full-time newspaper job without a lot of help. I offer sincere gratitude and a lifted glass to Elaine Maisner and the good people at the University of North Carolina Press for their belief that this project was a good idea and their astonishing faith that I was the right person to do it; to Paul Hletko of FEW Spirits, Lew Bryson, Robert F. Moss, Matthew Rowley, Kat Kinsman, Jennifer V. Cole, and Fred Minnick for inspiration, advice, interviews, wise counsel and hand-holding; to Bob Peters, Colleen Hughes, Brian Lorusso, and Kevin and Heather Gavagan for help with recipes; to Andrea Weigl, Debbie Moose, Robert and Camille Simmons, Martin Frobisher and Jeanne Grinstead, Tim Purvis and Laurie Purvis, Nathalie Dupree and Jack Bass, Regina and Doug Charboneau, and Dr. Jessica Harris for coming through with aid, comfort, and free lodging for a wayfaring writer; to Angel Postell and the hardworking crew of BevCon; to the staff at every distillery I visited, for their patience with my many, many questions; and most of all, to my husband, Wayne Hill, and our son, Chris Hill, who put up with a lot and who always have my back, even when I'm not working on a book.

INDEX

Baker, Richard, 86–87

Ball, Troy, 67–69

Ban vodka, 73

Banyan Reserve vodka, 148

Bardstown, Ky.: Bardstown Historical
Society, 121; Oscar Getz Museum of
Whiskey History, 121; Willett Dis-
tillery, 121, 123

Bardstown Historical Society, 121

Barley, 5, 14, 23, 30, 37, 40, 42, 72, 85,
140; malted barley, 33, 41, 57, 58, 71,
147, 176

Barrel aging/resting, 19, 32, 42, 117;
bourbon, 23; brandy, 37, 111–12; corn
whiskey, 23, 104; gin, 21, 22, 74, 78,
91, 115, 120, 124; poitín, 73; rum, 22,
70, 165–66; Tennessee whiskey, 23

Barrel House Distilling Co., 99 (map),
103–4

Barrels, 8, 15, 19, 22, 23, 32, 57, 67, 85,
104, 120, 141, 164, 175, 176

Barton 1792, 121

Bayou Rum, 172

Beam, Joseph, 112

Beam, Minor Case, 112

Beam, Paul, 112

Beam, Steve, 112

Bell, Amy, 114

Bell, Darek, 114–15

Belle Grove whiskey, 42

Belle Isle Craft Spirits, 43

Belle Meade whiskey, 116, 117, 126
(recipe)

Belmont, N.C.: Muddy River Distillery,
16, 50 (map), 69–71; Riverside
Marina, 69

Belmont Farm Distillery, 43

Benson, N.C.: Broadslab Distilling, 51
(map), 55–57

BevCon, 11–12

Beverage Tasting Institute, 174

Big Machine Platinum Vodka, 118

Birmingham, Ala.: Avondale Brewing
Company, 150; Avondale Spirits, 150;
Wooden Goat restaurant, 150

Black Coral Rum, 151

Blackwell, Scott, 12, 89–90, 93

Blairsville, Ga.: Granddaddy Mimms,
152

Bloomery Plantation, 1–2, 12, 27 (map),
38–40, 46 (recipe)

Bluegrass Distillers, 103, 122

Blue laws. *See* State and local
regulations

Blue Note Grill, 59

Blue Ridge Distilling Co., 73

Boar Creek whiskey, 44

Boehm, Jacob, 112

Bogdenovich, Mike, 145–47

Boiler Room restaurant, 58

Bondurant Brothers Distillery, 43

Boone, Daniel, 14

Boone County Distilling, 122

Bostic, N.C.: Blue Ridge Distilling Co.,
73

"Botanical-style" gin, 22, 59, 91

Bounces (flavored moonshines), 76

Bourbon, 2, 19, 57, 78, 85, 87, 91, 106,
112, 119, 145, 155; history, 5, 15, 102,
104, 121; standards for, 23

Bourbon producers: ASW Distillery,
130 (map), 136–38, 153 (recipe),
155; Barrel House Distilling Co., 99
(map), 103–4; Broadslab Distilling,
51 (map), 55–57; Cathead Distillery,
158 (map), 167, 168–69, 173 (recipe);
Charleston Distilling Co., 81 (map),
89, 93–94 (recipe); Copper Horse
Distilling, 80 (map), 86–87; Corsair Distillery, 98 (map), 101, 114–16;

Rougaroux rum, 164

Roulaison Distilling Co., 172

Roundstone rye, 37

Rum, 22, 31, 38, 58, 88, 104, 119, 125 (recipe), 142–43, 161–62, 165–66; history of in U.S., 162

Rum producers: Barrel House Distilling Co., 99 (map), 103–4; Broadslab Distilling, 51 (map), 55–57; Cajun Spirits Distillery, 172; Charboneau Distillery, 158 (map), 169–71, 173 (recipe); Chesapeake Bay Distillery, 44; Copper Horse Distilling, 80 (map), 86–87; Daufuskie Island Rum Company, 92; Donner-Peltier Distillers, 158 (map), 164–66; Firefly Distilling/ Spirits, 83, 92; Fish Hawk Spirits, 131 (map), 133, 145–47; H&H Distillery, 73; Independent Distilling Company, 152; James River Distillery, 27 (map), 33–36, 46 (recipe); John Emerald Distilling Company, 130 (map), 134, 140–41, 153 (recipe); Louisiana Spirits Distillery, 172; Lula Restaurant Distillery, 172; Mother Earth Spirits, 51 (map), 57–58; Mt. Defiance Cidery & Distillery, 4, 27 (map), 29, 37–38; Muddy River Distillery, 16, 50 (map), 69–71; Old Glory Distilling Co., 124; Old New Orleans Rum Distillery, 158 (map), 161, 162, 163–64; Outer Banks Distilling, 74; Prichard's Distillery, 98 (map), 118–19; Rank Wildcat Spirits, 172; Rich Grain Distilling Co., 158 (map), 167–68; Richland Rum, 130 (map), 141–43; Rollins Distillery, 151; St. Petersburg Distillery, 131 (map), 148; South Florida Distillers, 151; Striped Pig Distillery, 81 (map), 84 (ill.), 88–89,

93 (recipe); Vitae Spirits, 44; Wicked Dolphin Artisan Rum, 151; Williamsburg Distillery, 44

Russell, Jimmy, 87

Rye: Canadian, 23; grain, 3, 14, 22, 23, 30, 36, 41, 42, 84, 89, 91; whiskey, 14, 23, 30, 32, 37, 42, 45 (recipe), 117, 155

Ryemageddon rye, 115

Rye producers: Catoctin Creek, 27 (map), 29, 36–37; Charleston Distilling Co., 81 (map), 89, 93–94 (recipe); Corsair Distillery, 16, 98 (map), 101, 114–16; Doc Porter's Craft Spirits, 50 (map), 71–72; George Washington's Distillery, 15, 27 (map), 30–33; High Wire Distilling Co., 21, 81 (map), 89–91, 126 (recipe); Kentucky Artisan Distillery, 122; Kentucky Peerless Distilling Co., 99 (map), 108–9; Lazy Guy Distillery, 130 (map), 134 (ill.), 135–36, 154 (recipe); Limestone Branch Distillery, 98 (map), 112–13; Mt. Defiance Cidery & Distillery, 4, 27 (map), 29, 37–38; Old Pogue Distillery, 99 (map), 102 (ill.), 104–7; Prichard's Distillery, 98 (map), 118–19; Reservoir Distillery, 27 (map), 36; Silverback Distillery, 44; Thirteenth Colony Distilleries, 130 (map), 144–45. See also Whiskey producers

St. Augustine, Fla.: Ice Plant restaurant, 150; St. Augustine Distillery, 131 (map), 149–50

St. Augustine Distillery, 131 (map), 149–50

St. Petersburg, Fla.: St. Petersburg Distillery, 131 (map), 148

St. Petersburg Distillery, 131 (map), 148